Mathematical Beginnings

Problem Solving for Young Children

By Noel Graham and Janine Blinko

Illustrated by Mike Spoor

(additional illustrations by Lisa Spurr)

Claire Publications

First published 1988
Second impression 1990

Published by
Claire Publications
York House
Bacons Lane
Chappel
Colchester
Essex CO6 2EB

ISBN 1 871098 00 9

Printed in Hong Kong by Centurion Press Ltd.

This book has been prepared on an Atari ST, using Publishing Partner software by Softlogic and printed on a Hewlet Packard Laserjet laser printer.

We would like to thank SDL (UK) Ltd. for their invaluable help and assistance in supplying the machines and software and their advice in general.

Typeset from disk by The Moving Finger Company, Southampton (0703) 229041

TABLE OF CONTENTS

FOREWORD

The mathematics lesson has as much to do with learning to learn as with learning mathematics. Mathematical fact acquired by active, enthusiastic pupil participation creates children who are able to think for themselves. All the activities in this book encourage pupils to think for themselves and develop their own solutions to problems.

The ideas in this book are aimed at developing children's knowledge and understanding of mathematics by involving them in a wide variety of enjoyable and stimulating problem solving activities. Guidelines for the teacher and extension challenges are provided with each activity. The reproducible worksheets are intended as a trigger, a creative beginning to a mathematical exploration. Completing them is not an end in itself, they are a starting point, a means of helping children record their work and an aid for the teacher preparing materials. We hope that the users - pupils and teachers - develop the habit of asking questions which take the ideas further. The open ended nature of most of the activities enables them to be used in subsequent years with the same children.

To help children with their recording the reading skills necessary to complete the worksheets have been kept to a minimum.

We believe that the only purpose in learning mathematics is to solve problems. Children can only become good 'problem solvers' if they are involved in new, unusual and stimulating situations that challenge them. The situations may appear trivial, lead to developments of mind shattering importance or make us laugh or cry. Everything we want children to learn can be presented as a problem. If for example we want children to learn number bonds so that they can rapidly recall them, they need numerous opportunities to use them in enjoyable and meaningful ways. The child who has explored what happens when numbers are combined in a variety of ways is likely to understand and retain the number facts essential to the manipulation of larger numbers.

Expecting children to record their mathematics serves a number of useful functions. It helps clarify and reveal mathematical ideas. If the problem is complex, recording data as it emerges helps find a solution. Communication through mathematics gives a child power over it, consequently exploring a variety of methods of recording helps develop that power. Finally and realistically we need a record of childrens' work, to show what they have achieved.

There are three forms of classroom display necessary in the teaching of mathematics: permanent, semi-permanent and transitory. The first consists of reference material for the children e.g. number lines and squares placed at a level children can see, touch and use. The semi-permanent displays are usually of childrens' work, mounted with care to show how much we respect their efforts. They can also be materials displayed by the teacher to beautify the room or as a starting point for an investigation. Finally, there are the displays that remain for an hour, a day or even a week. These are a record of ongoing activities recorded on the chalkboard or large sheets of paper. The creative use of these three forms of display generates the atmosphere in the class. How barren a classroom can look without them!

The activities which follow assume that equipment is readily available to pupils at all times.

INTERLOCKING CUBES

The following ideas are designed for use with interlocking cubes.
It is assumed that children will have played freely with the cubes before any structured activities begin.

CAMILLA CAMEL

Cube Sheet 1

****

These activities build on children's free play experiences.

****

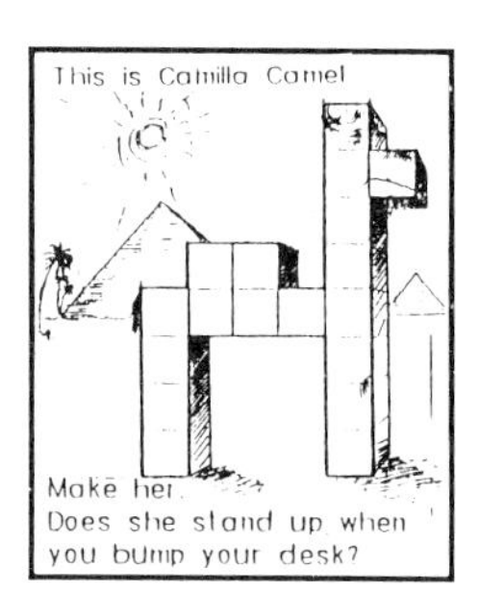

If Camilla falls over, ask the children to invent ways of making her more stable. Any solution is acceptable.
There is great value in encouraging the children to discuss their strategies, and asking them to explain why they have been successful.

*** A Challenge ***

Make a multicoloured bird that will not fall over when the desk is bumped

Extensions

1. Make a zoo.
2. Make an animal family.
3. Use the Three Bears Activity in the Cuisenaire section of this book. Ask children to make bears which do not fall over.

This is Camilla Camel

Make her.
Does she stand up when you bump your desk?

Cube Sheet 1

CONSTRUCTIONS

Cube Sheets 2,3 and 4

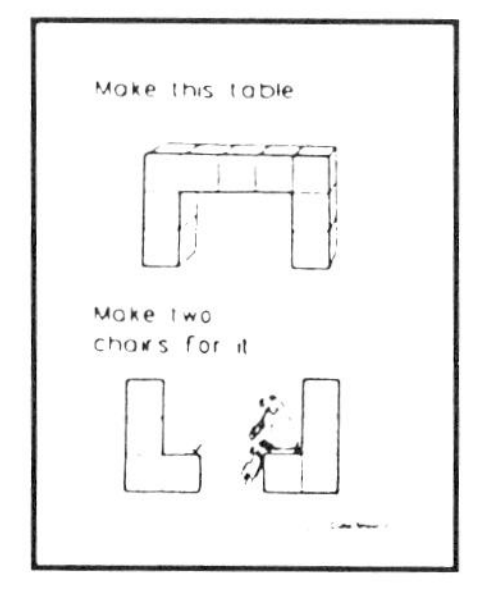

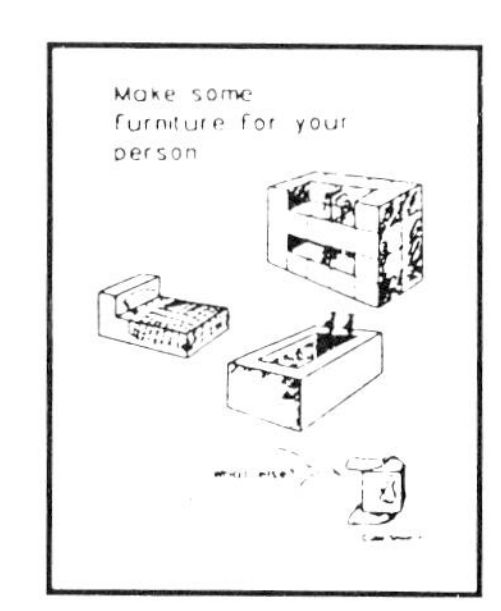

****

These sheets are starting points for 3 dimensional work. Other starting points may include constructing a playground, vehicles or boats to hold different numbers of passengers or quantities of freight etc.

****

1. Cube sheets 2 and 3 are designed to lead children from 2 dimensional models of real objects to 3 dimensional ones.
   This should lead to discussion about different kinds of furniture for different uses.

2. On cube sheet 4, children choose an item of furniture and guess how many cubes they need to make it.

3. When it is made, discuss how accurate their estimate was, and if the model is the right size for their person.

4. More furniture should be made with similar challenges.

<u>Extensions</u>

1. Price the cubes, for example 1p each <u>OR</u> by colour; 1p for red, 2p for green etc. Children then price the furniture they have made.

2. Furnish a two-storey house for £5.00.

3. All the activiites discussed earlier can be repeated with differing financial constraints. For example " Make the biggest, most stable camel you can for £1.00", or "How much did Daddy Bear cost to make?"

4. Make a chair big enough for one of the children to sit on.

Make this table.

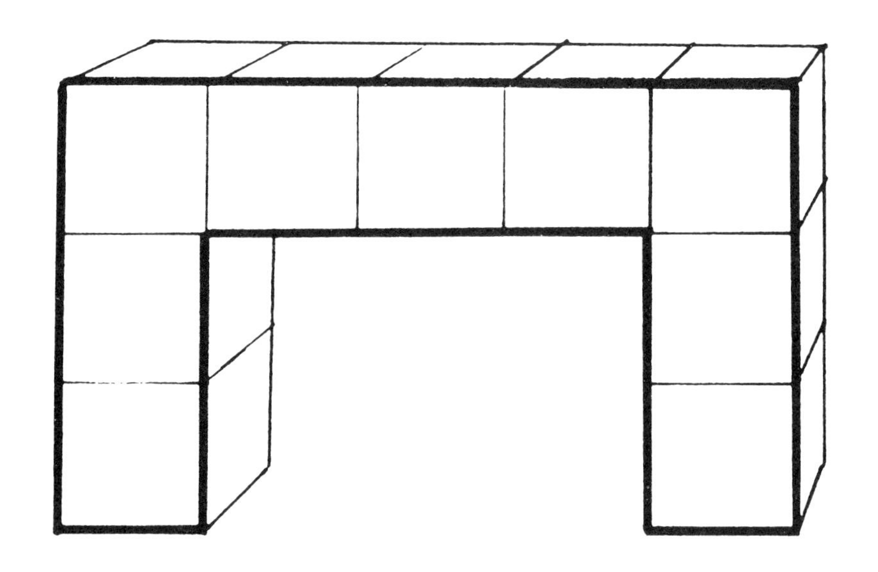

Make two
chairs for it.

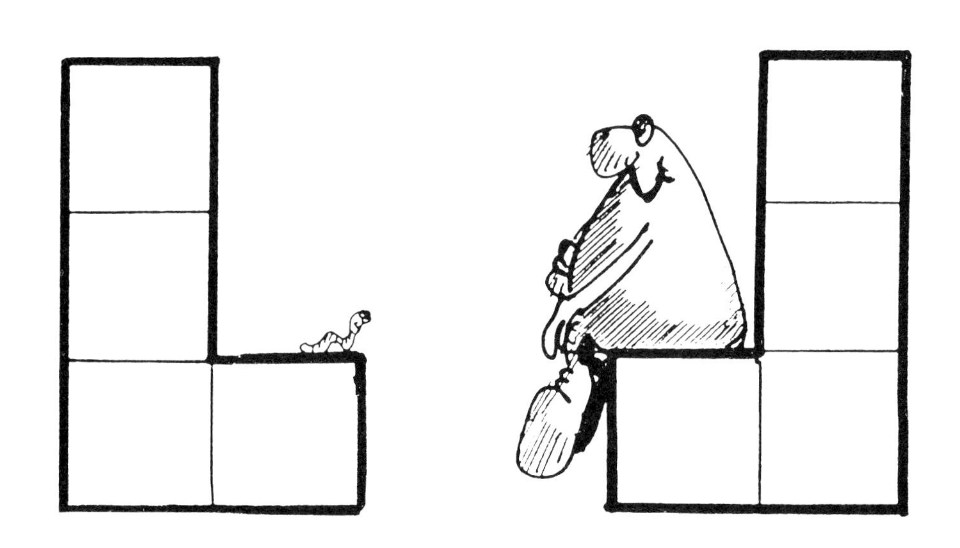

Cube Sheet 2

Change the table.

Change the chairs to match.
Make someone to sit on the chairs.

Cube Sheet 3

Make some furniture for your person:

Cube Sheet 4

DELIA DOG AND ROBERT ROBOT

Cube sheets 5 - 14

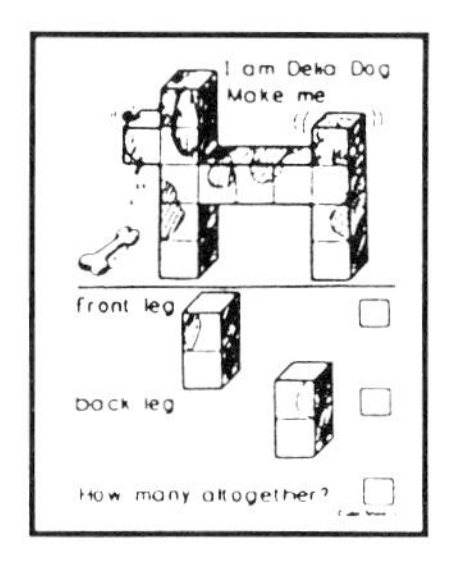

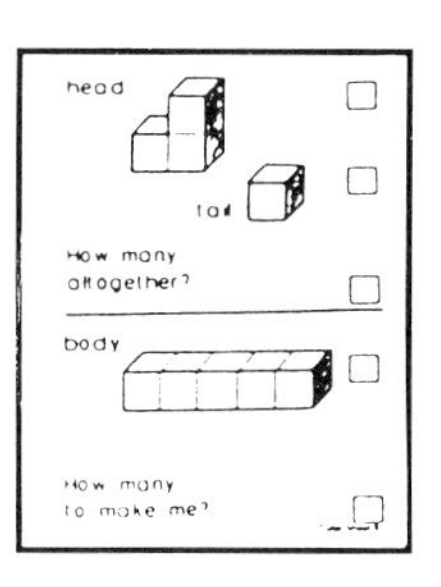

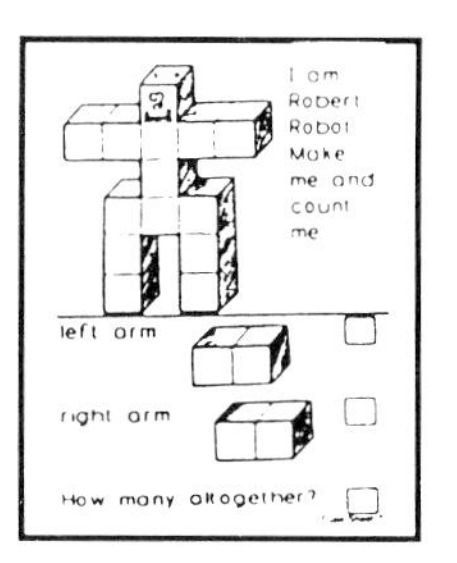

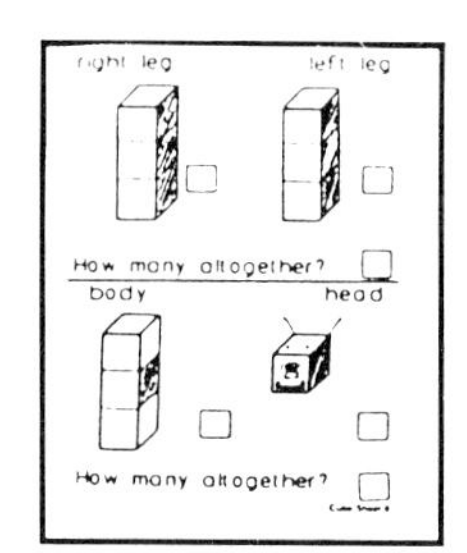

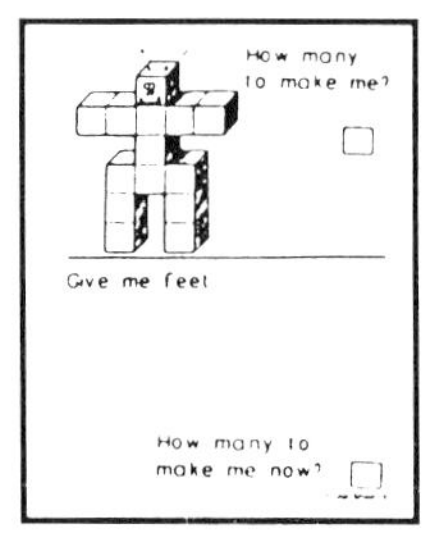

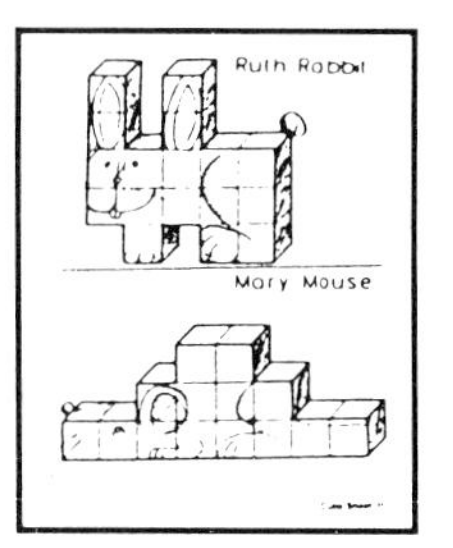

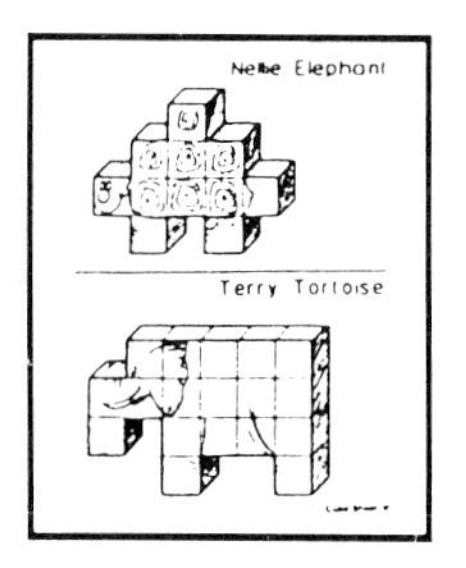

****

These sheets give children experience of counting and calculating. During the discussions of the activities opportunities will occur to formally introduce the concepts of multiplication and division as well as addition and subtraction. For example, "if one of Delia's legs uses 2 cubes how many cubes do you need for 4 legs?"

****

1. Sheets 5 to 9 are designed to be made into booklets. These explore ways in which numbers can be broken down.

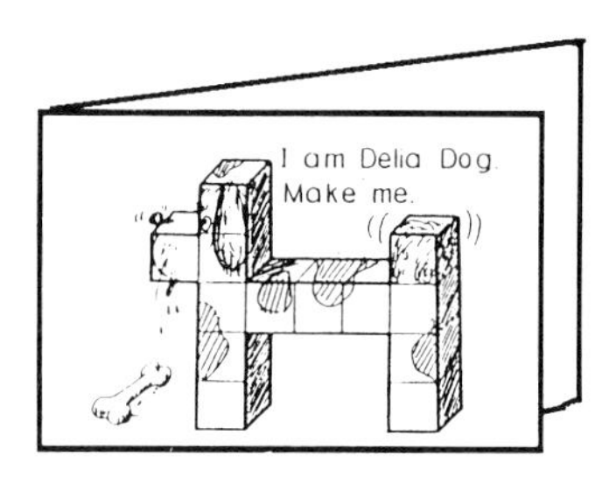

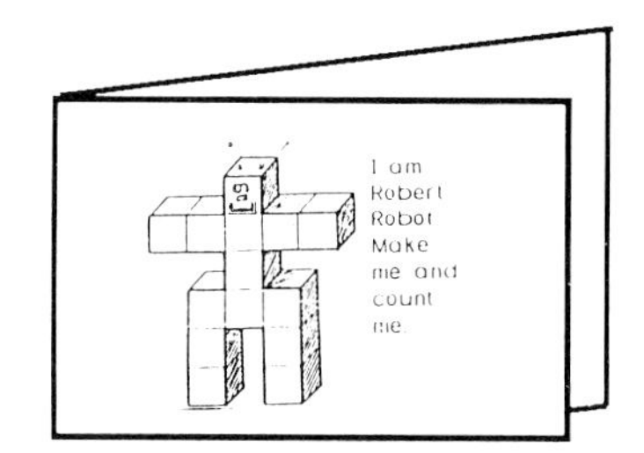

2. Sheet 10 gives plenty of scope for the development of Mathematical language, with suggestions like....
   - Make something taller
   - Make something wider etc.

3. Sheets 13 and 14 can be made into workcards or small worksheets for the children to make their own booklets.

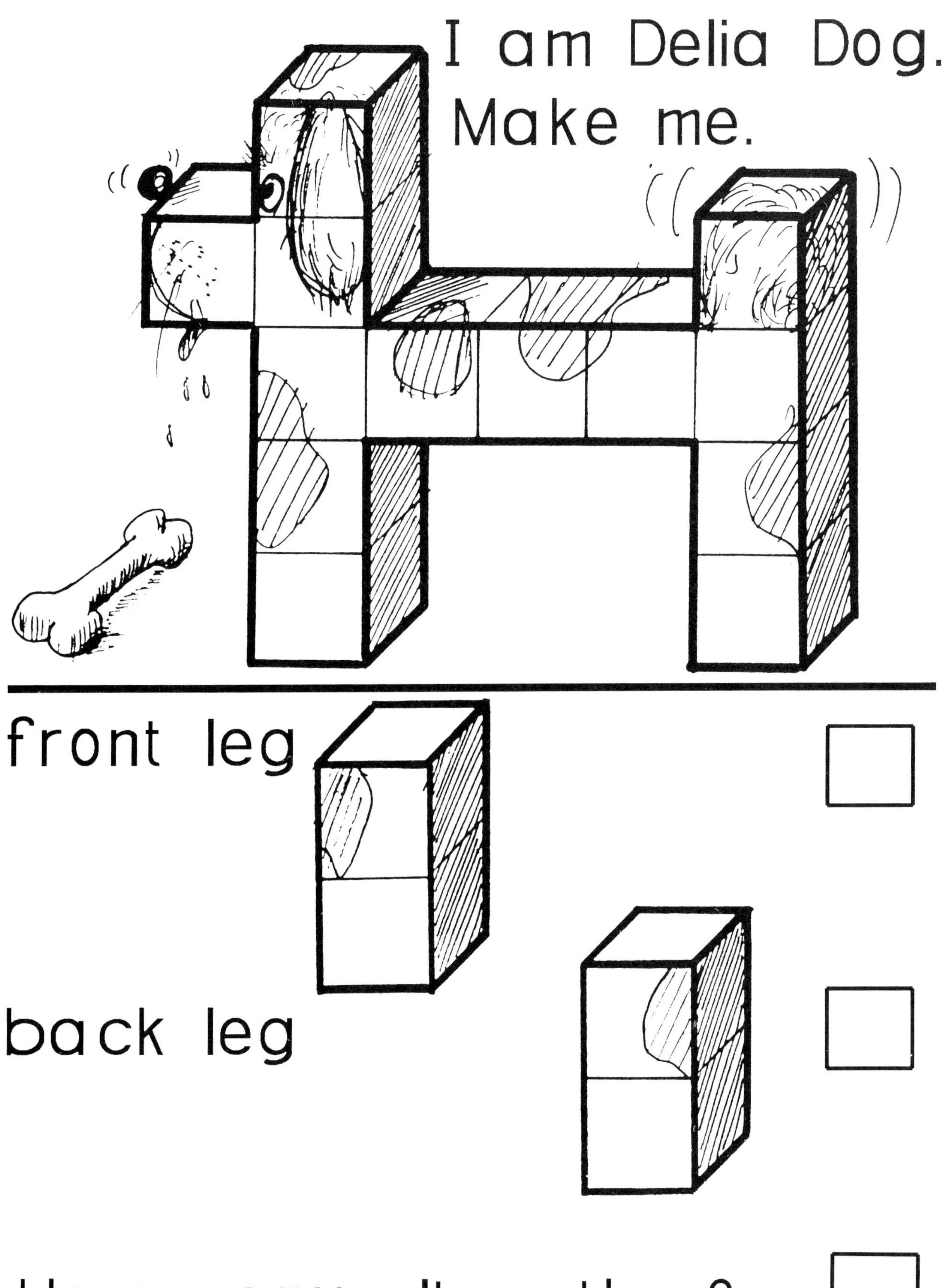

How many altogether?

Cube Sheet 5

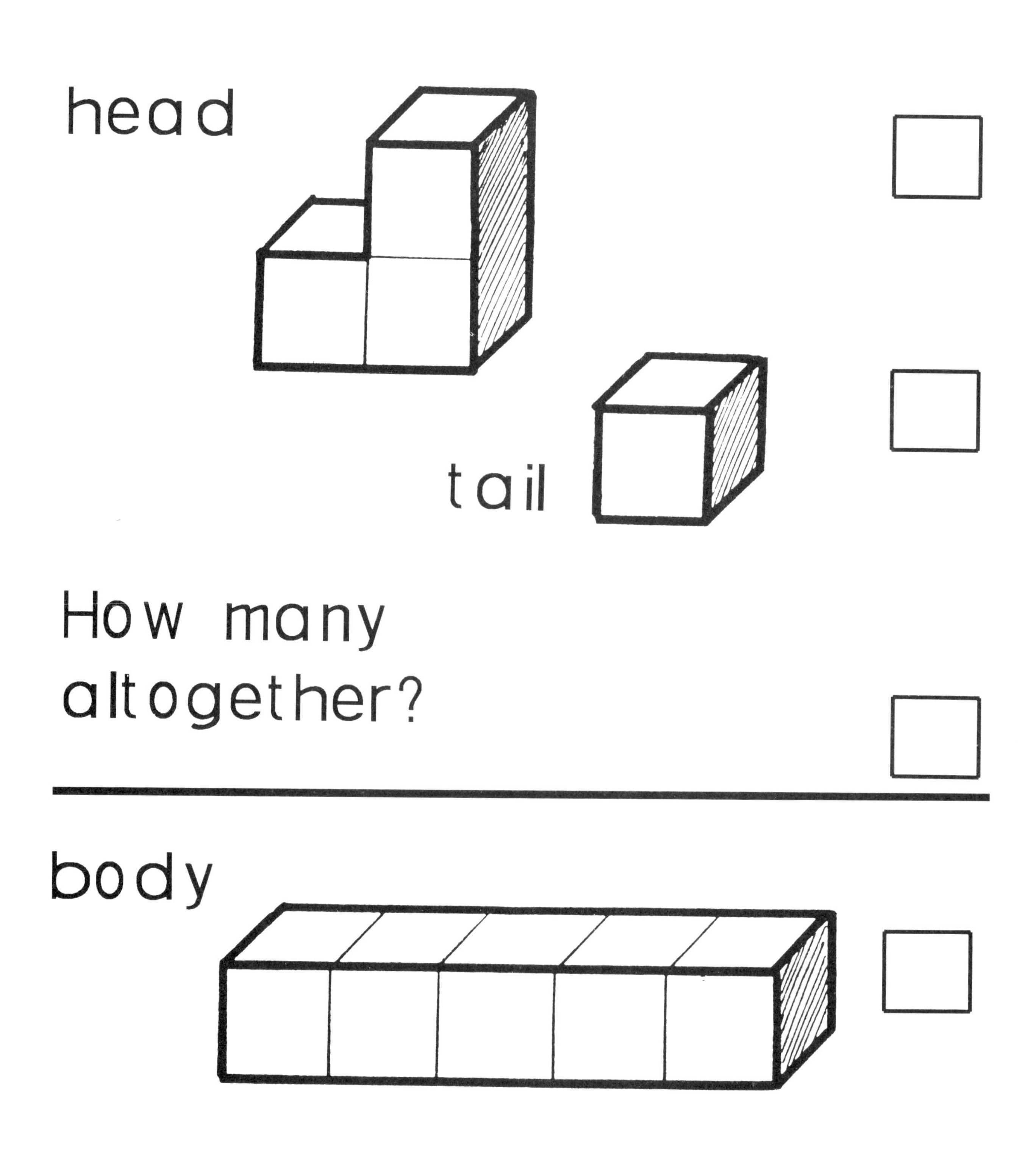

How many to make me?

Cube Sheet 6

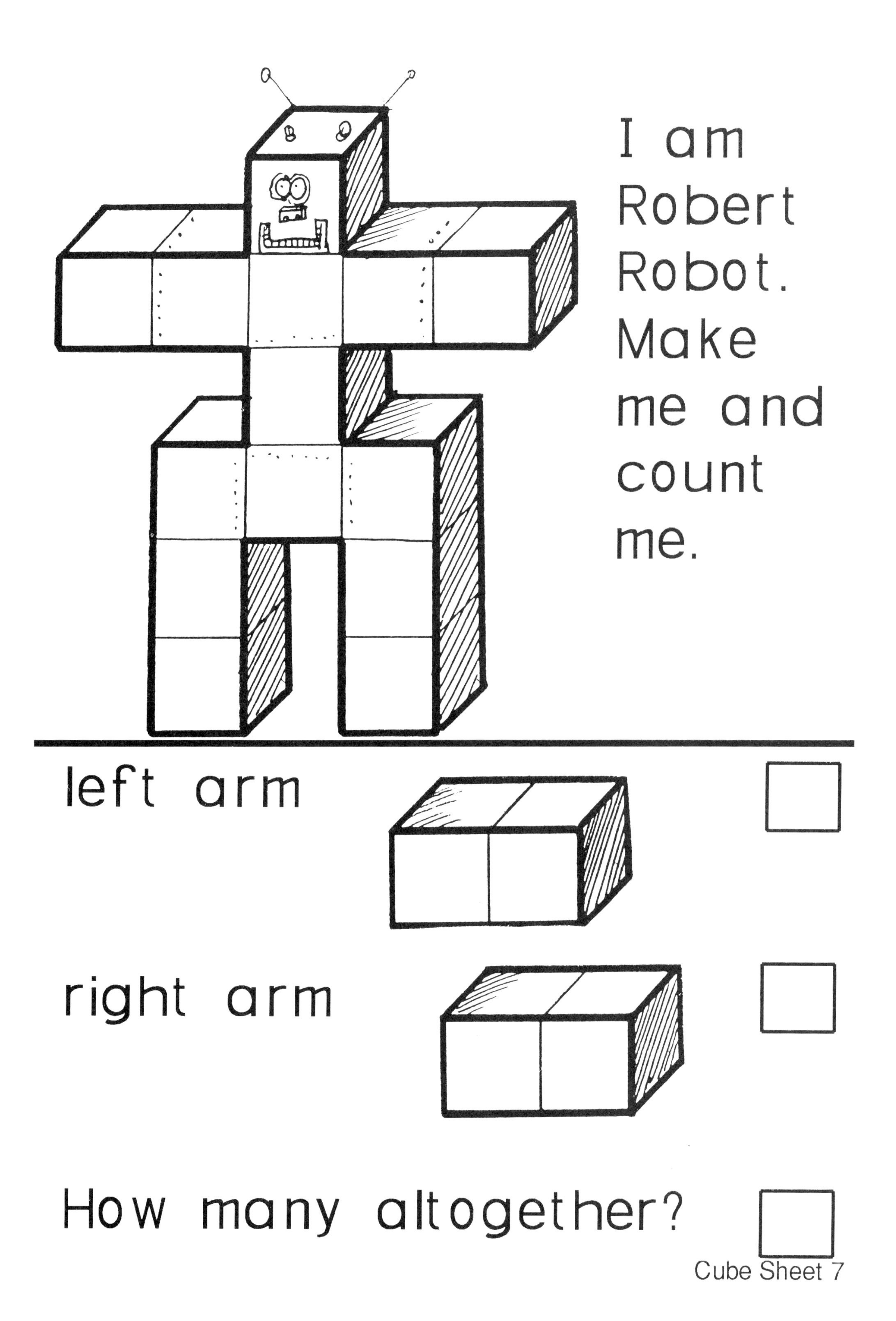
I am
Robert
Robot.
Make
me and
count
me.
left arm
right arm
How many altogether?
Cube Sheet 7

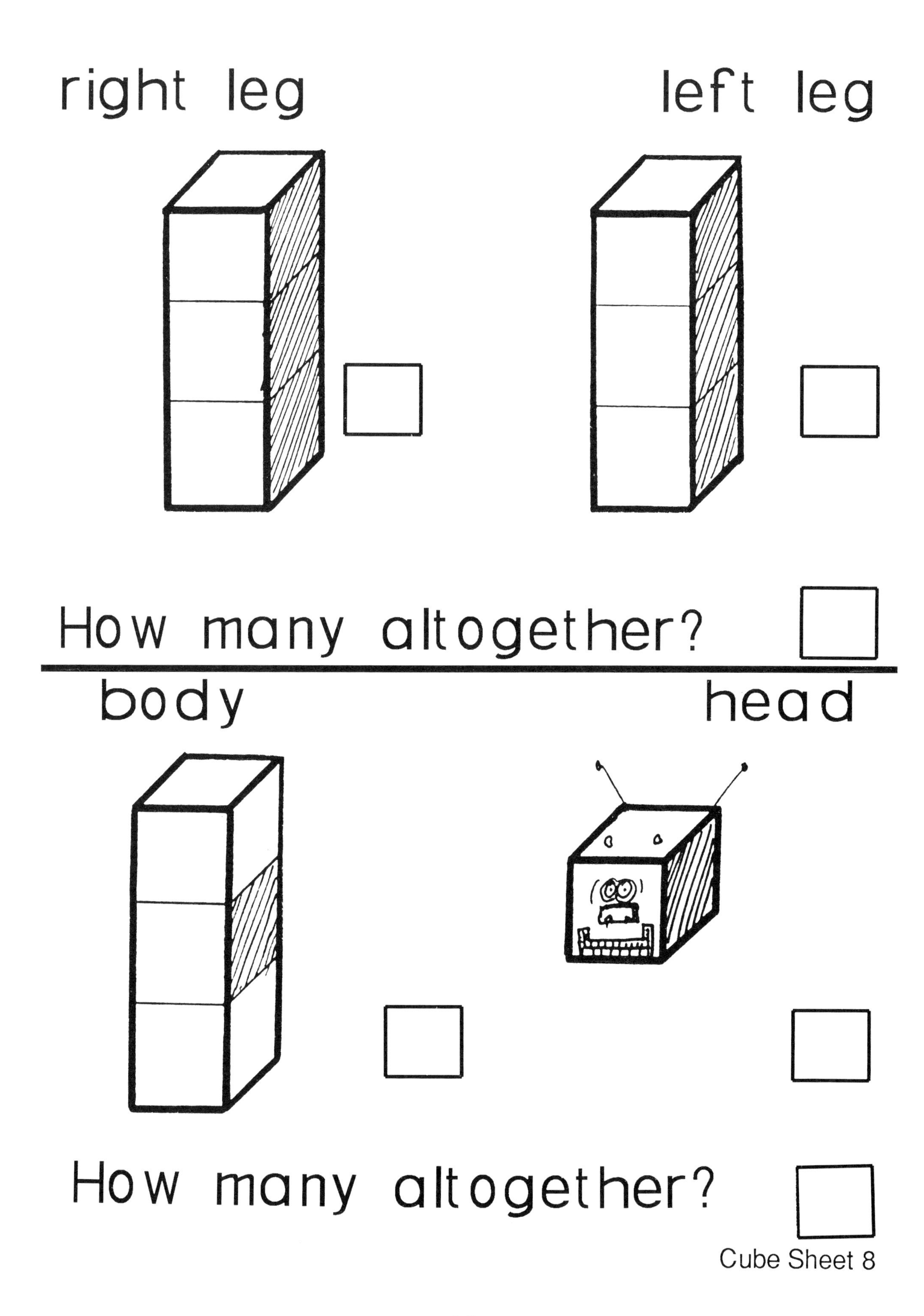
right leg
left leg
How many altogether?
body
head
How many altogether?
Cube Sheet 8

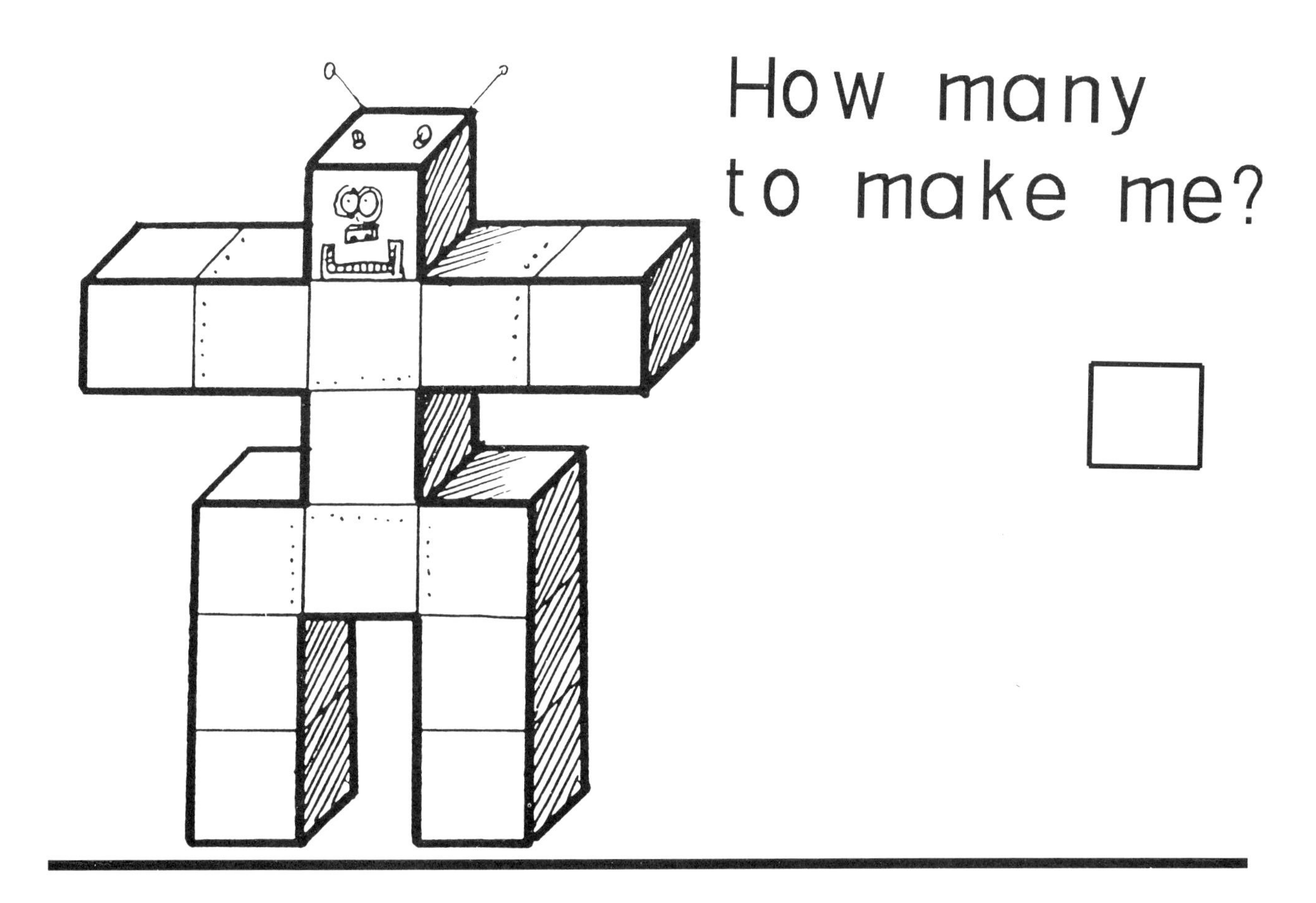

How many
to make me?

Give me feet.

How many to
make me now?

Cube Sheet 9

Take 10 cubes.

Use them all to make something

____

Take 10 more.

Make something different...
How are they different?

Cube Sheet 10

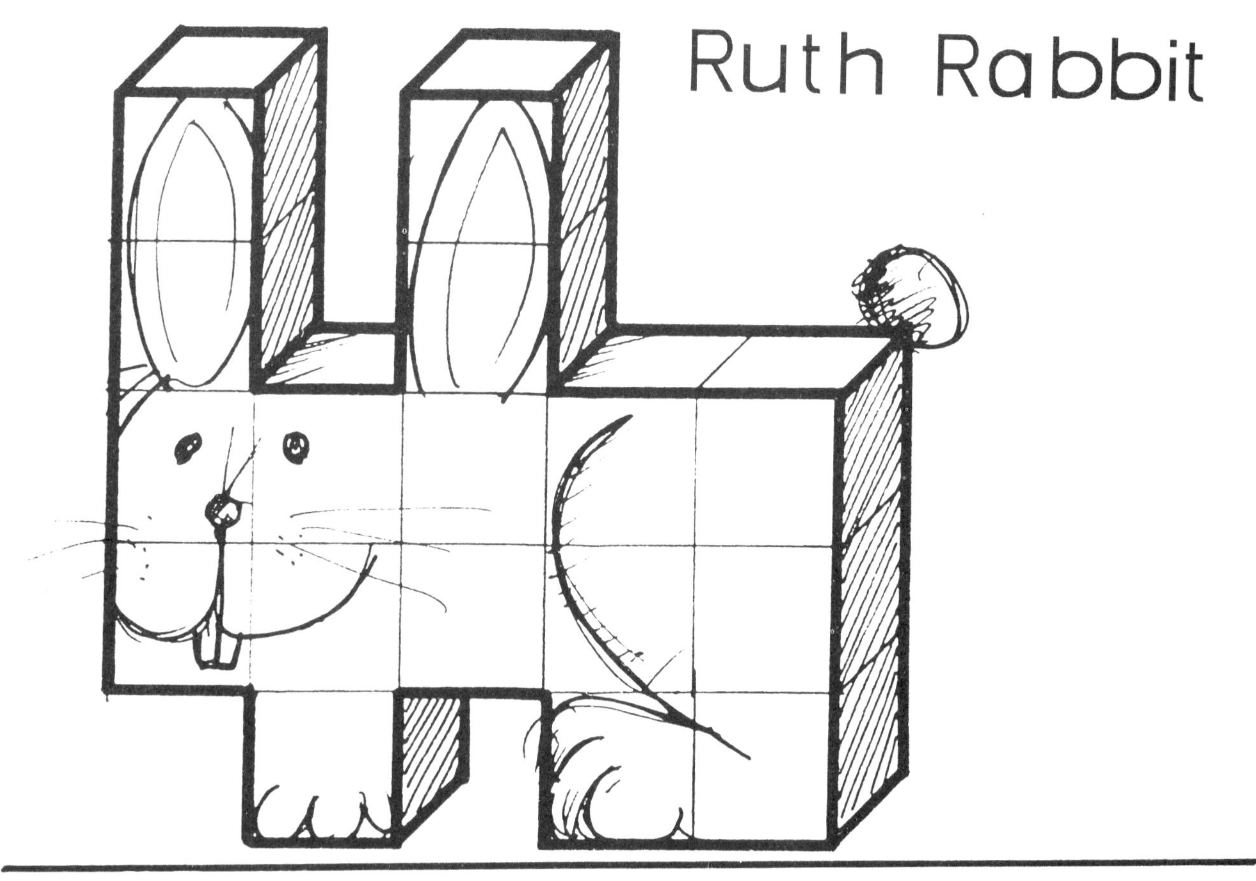

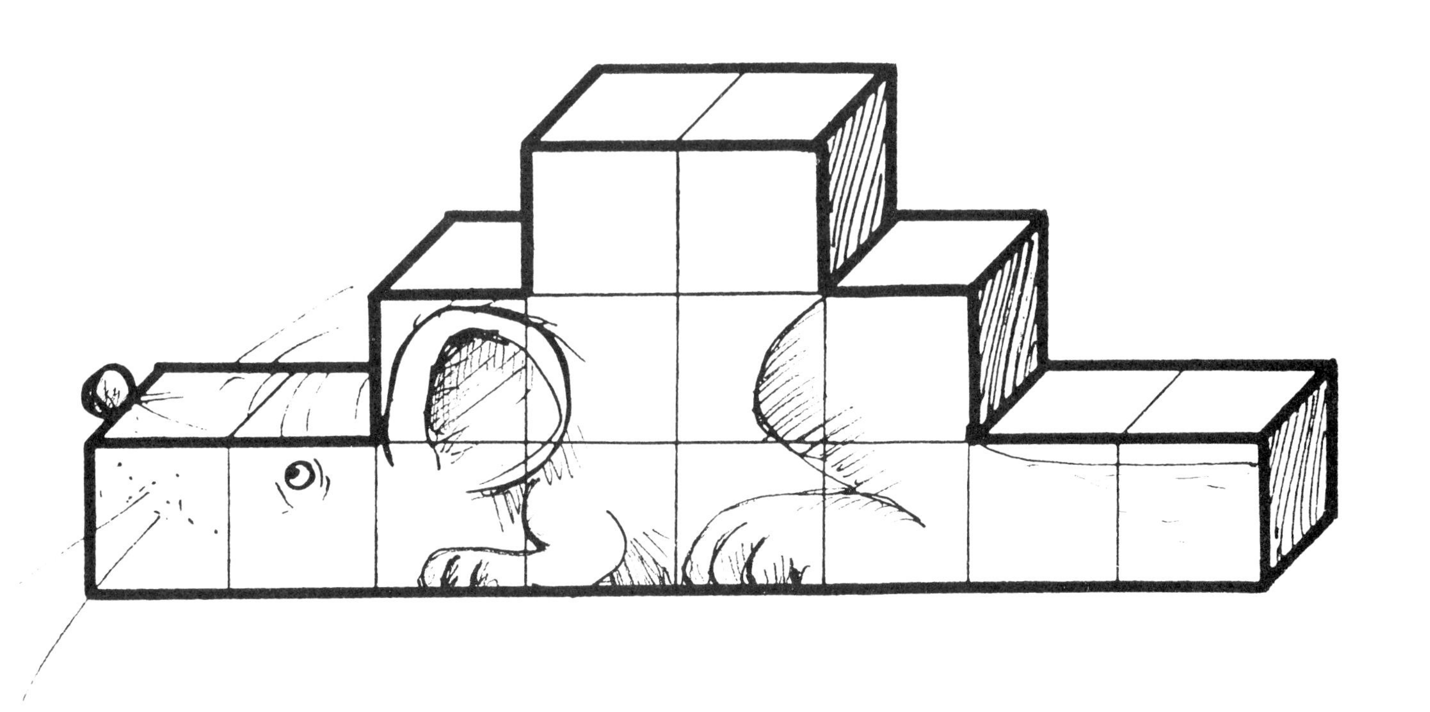

Cube Sheet 11

Terry Tortoise

Nellie Elephant

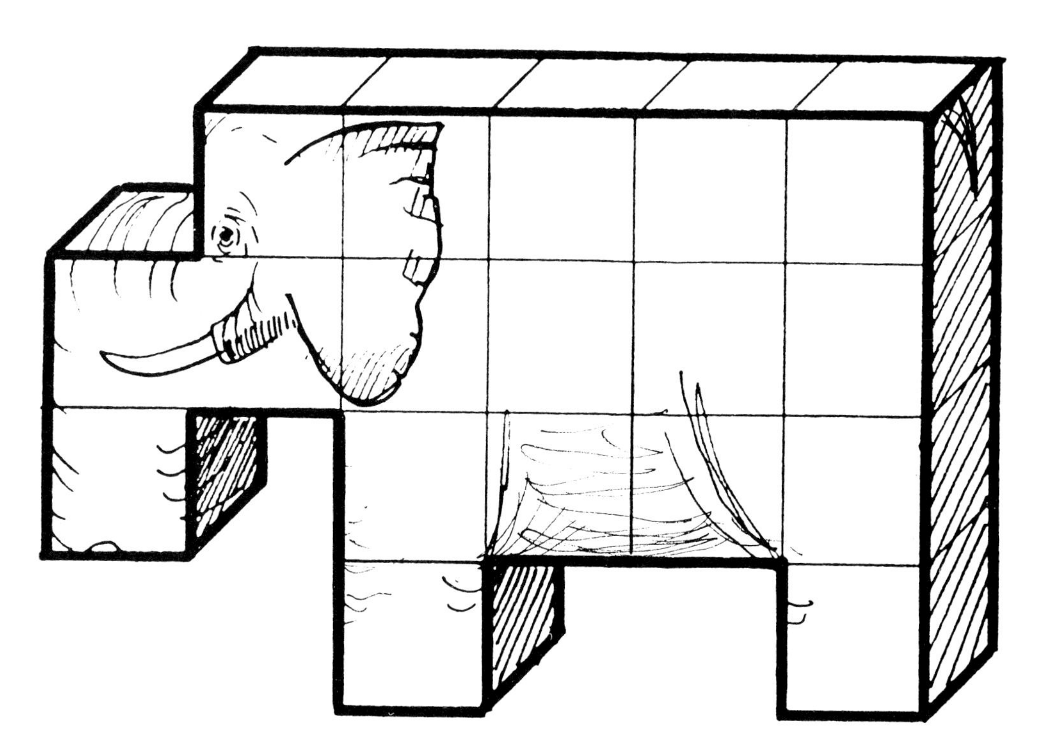

Cube Sheet 12

Sally Snake

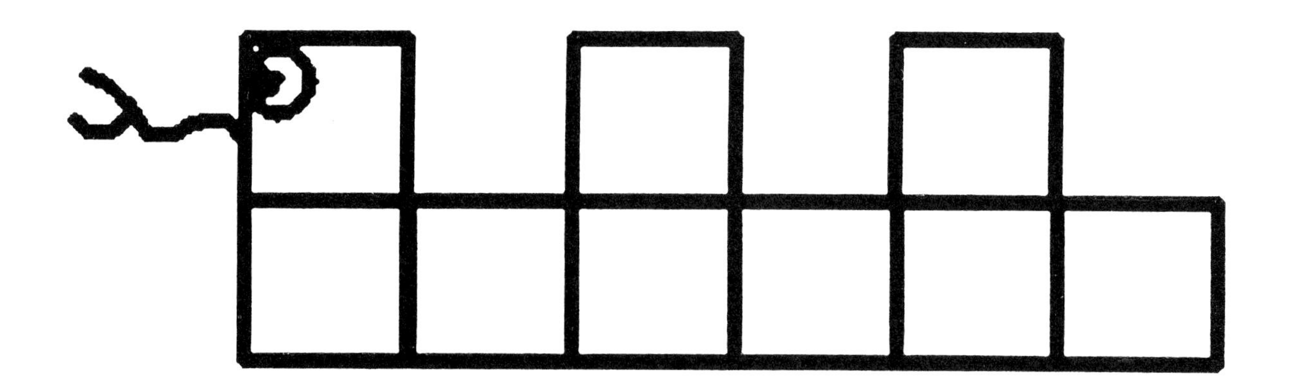

Penny Pyramid

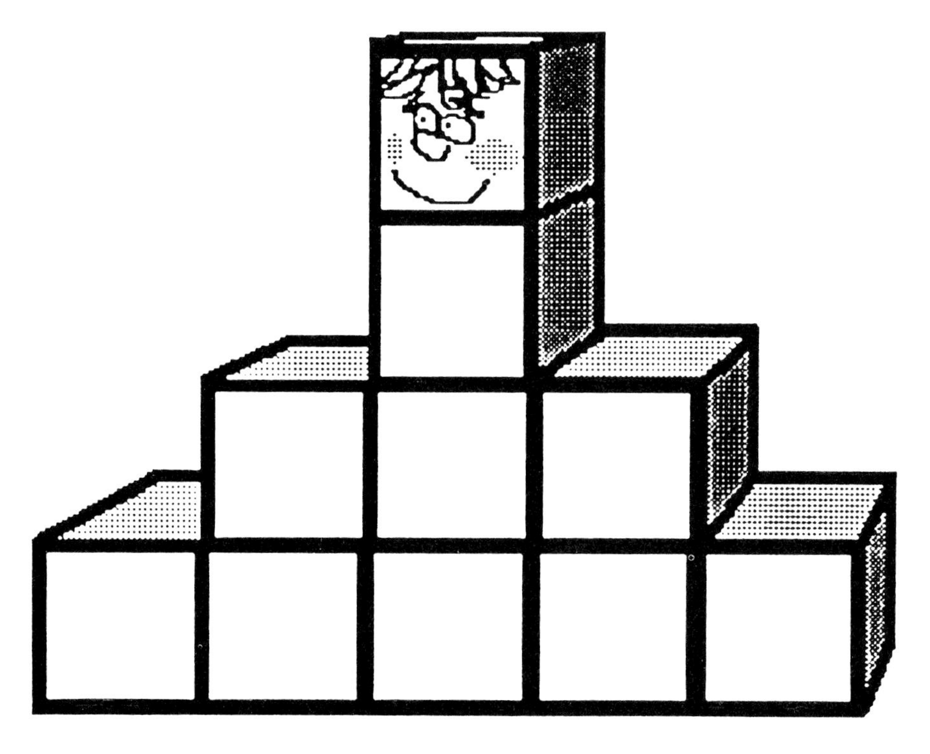

Cube Sheet 13

Cassie Cow

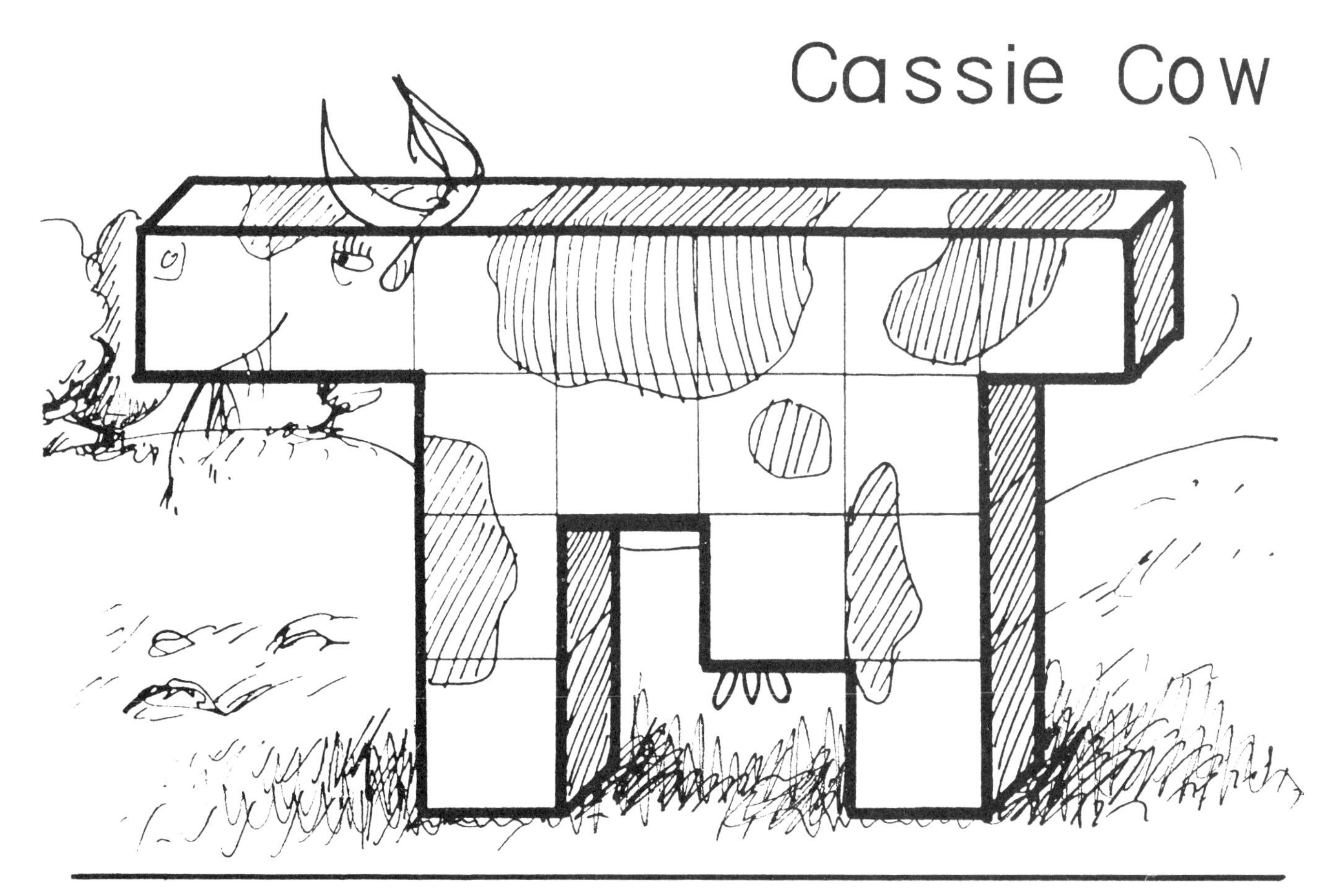

Simon Snail

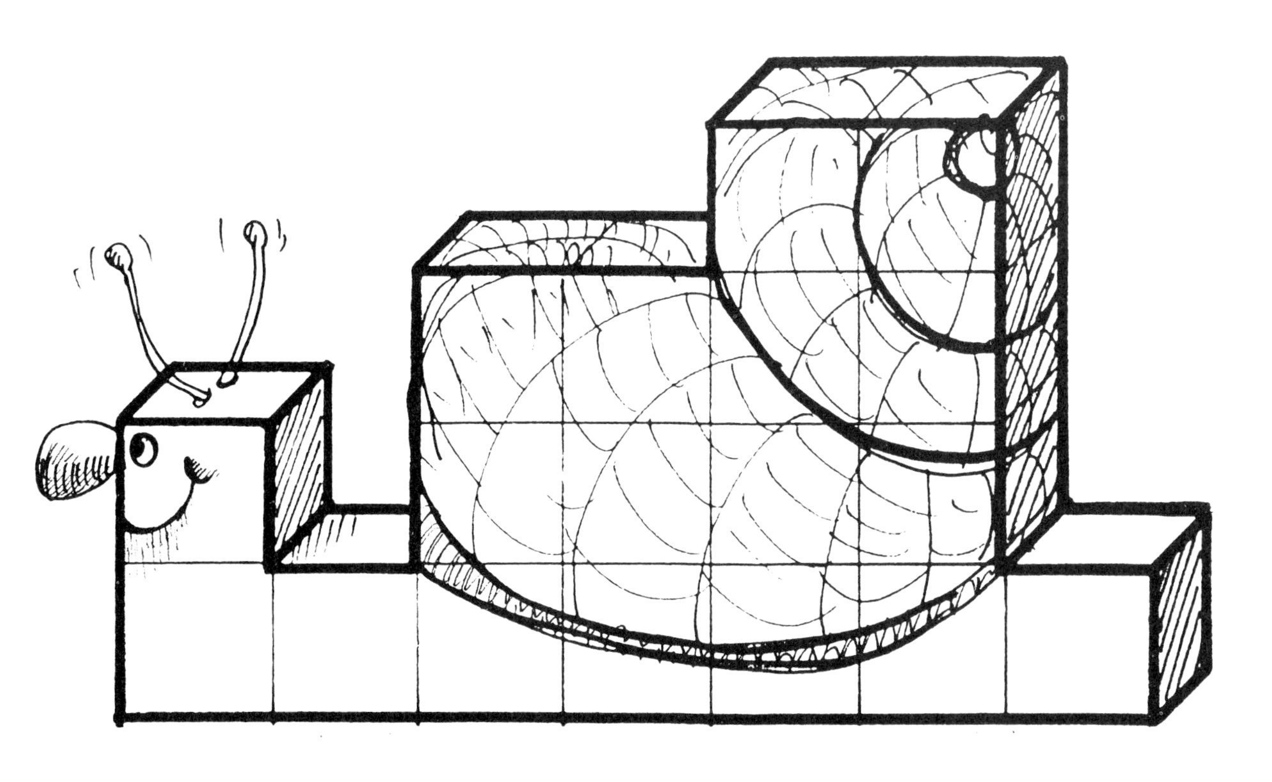

Cube Sheet 14

ESTIMATION

Cube Sheet 15

****

This activity encourages children to estimate, count and check

****

Ask the children to:-

1. Guess how many squares there are.
2. Count them.

They will have a variety of answers.

Record these and ask:-

i. Why are some answers different?
ii. How can we find out which answer is correct? (Encourage children to use cubes to help them count.)
iii. Can the cubes be arranged so they can be counted easily?

For example,

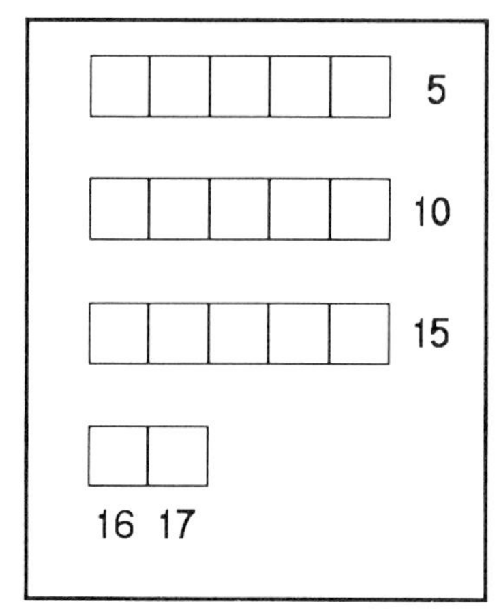

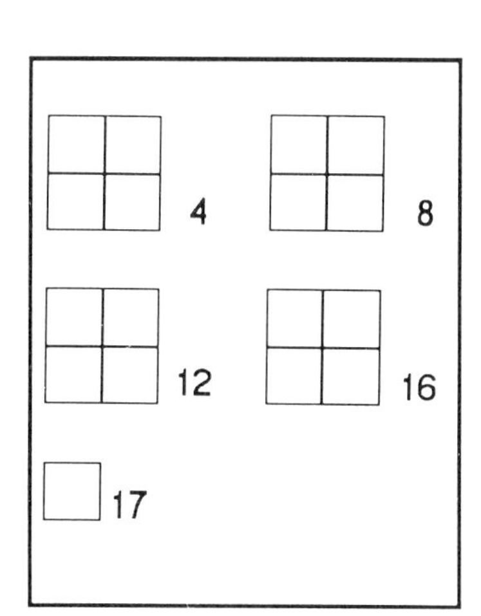

cont.....

Extensions

1. Cover the squares with cubes.

   Use them to make models.

2. Invent number sentences about the models.

For example,

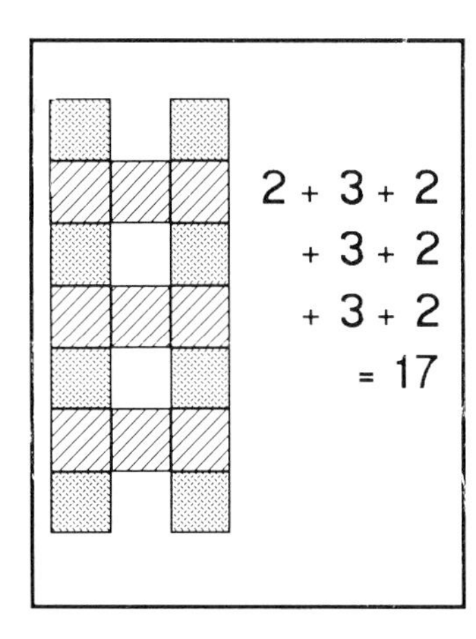

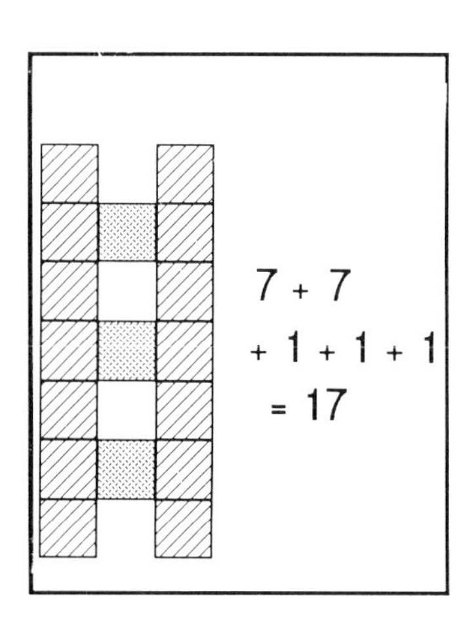

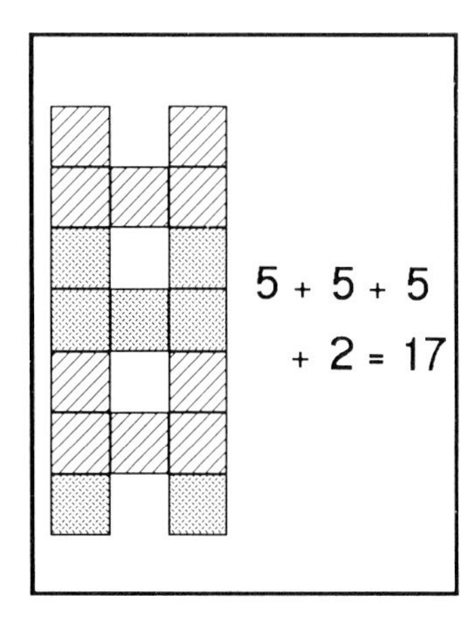

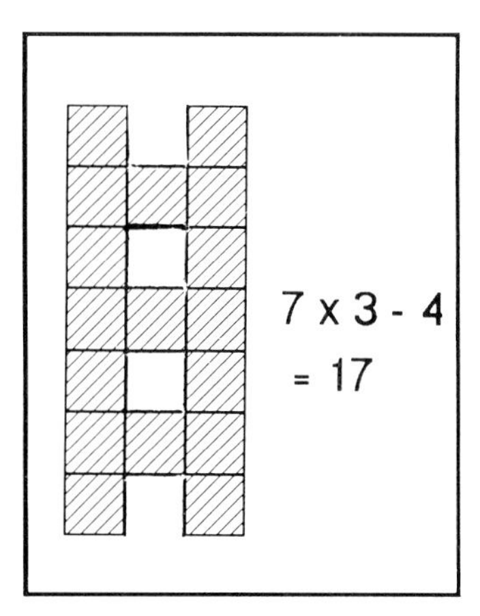

.... *It is vital the children 'see' and record their own number sentences.*

3. The idea can be reversed. Give the children number sentences and ask them to build a model from them.

4. This worksheet can be easily adapted for larger or smaller numbers by adding or removing squares.

Cube Sheet 15

TOWN PLANNING

Cube sheets 16,17 and 18

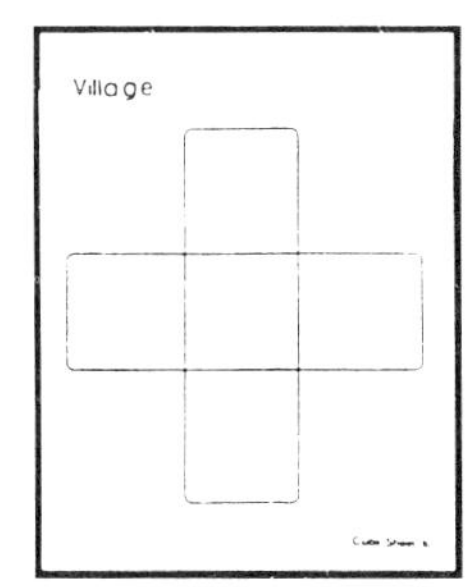

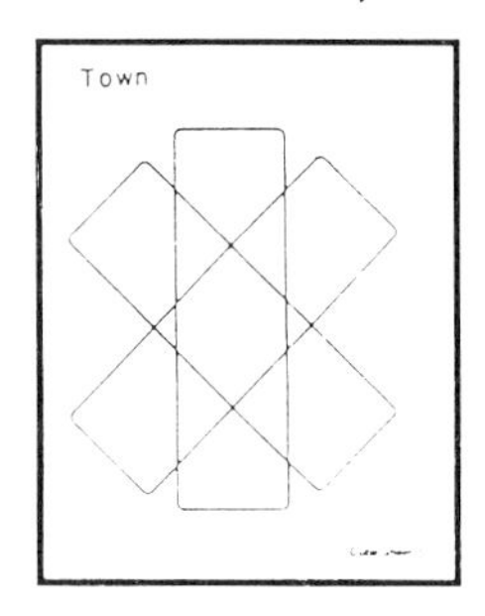

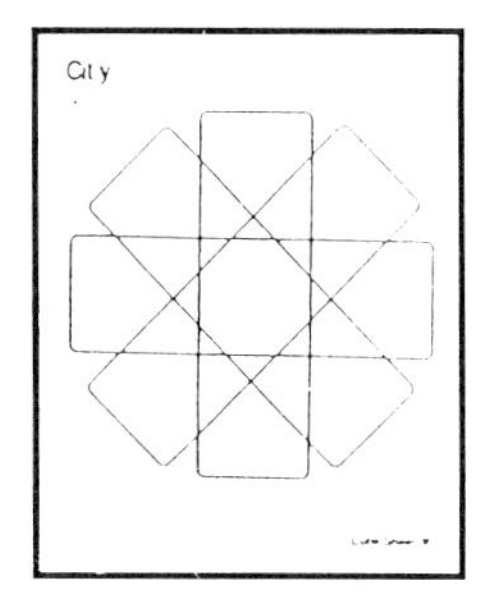

We recommend that these worksheets are enlarged.

****

These ideas develop children's logical thinking and spatial awareness, as well as their use of number and set theory. Children may find it helpful if teachers outline the regions in the appropriate colours.

****

In this activity children are creating villages, towns and cities which are made up of different coloured regions. There are red buildings in the red regions and blue buildings in the blue regions etc.

When two regions intersect the separate sections are called zones.

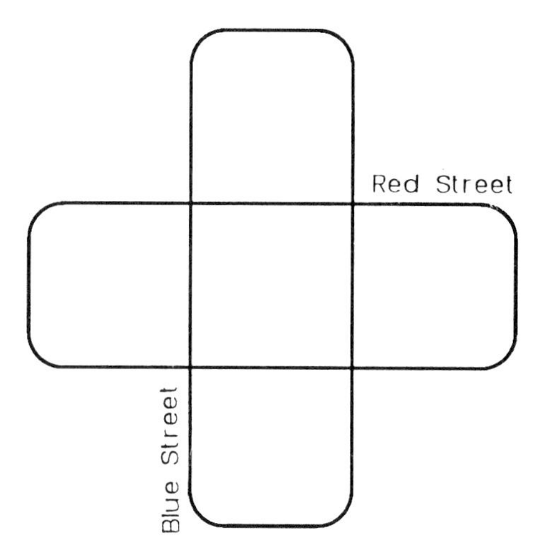

Any zone in the red region must have one red building on it, similarly for blue. For example in the village above there are three red zones and three blue zones.

Ask the children to decide what happens at the intersection. Encourage them to build upwards to make 2-storey buildings. In towns or cities where more regions intersect some buildings will be taller.

cont....

Questions

1. Which plot has the tallest building?

2. Do any plots have buildings of the same size?

3. How many cubes have been used altogether?

Extensions

Reverse the rules.

A red building may not be built in a red region, nor a blue building in a blue region. They should be built everywhere else that is possible.

This idea can be extended to the town and the city.

Further Questions

1. Explain how the two cities differ.

2. Which used the most cubes?

3. Give each building a value eg. Red is 1p
   Blue is 2p
   Yellow is 3p
   Green is 4p.

   Which building costs the most to build?

   Do any two buildings cost the same?

   Are there any buildings worth the same in both types of city?

Village

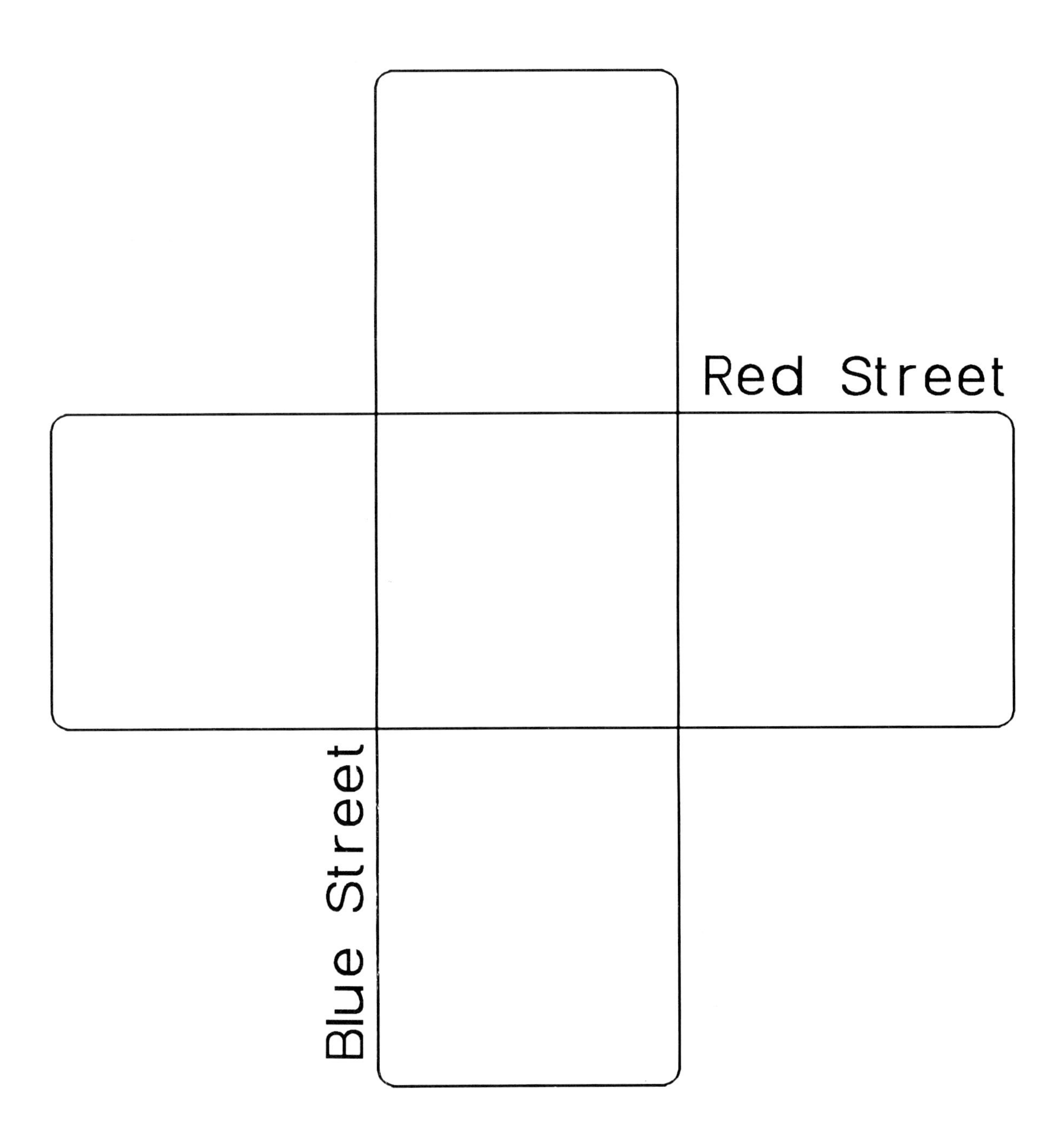

Cube Sheet 16

Town

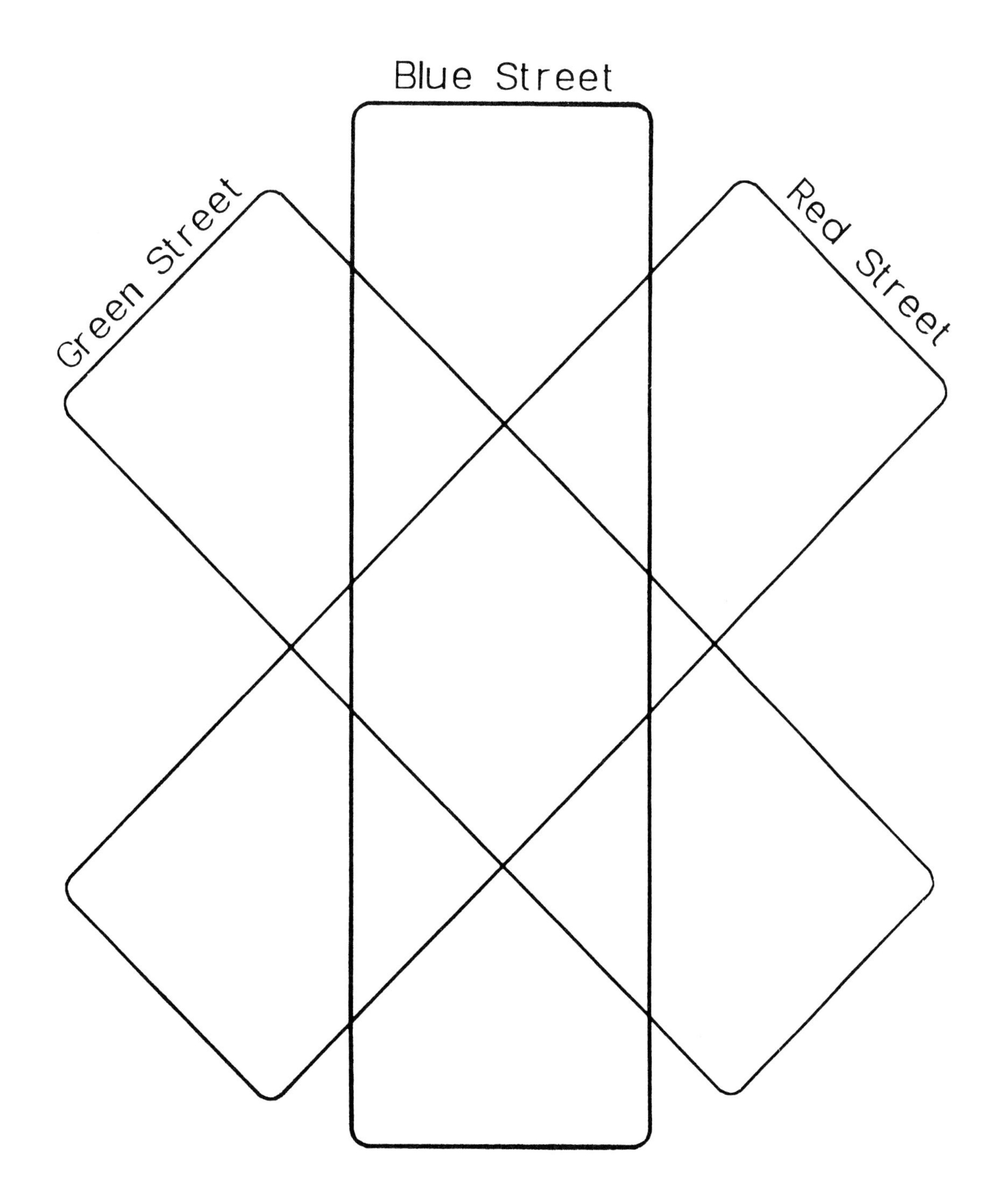

Cube Sheet 17

Cit y

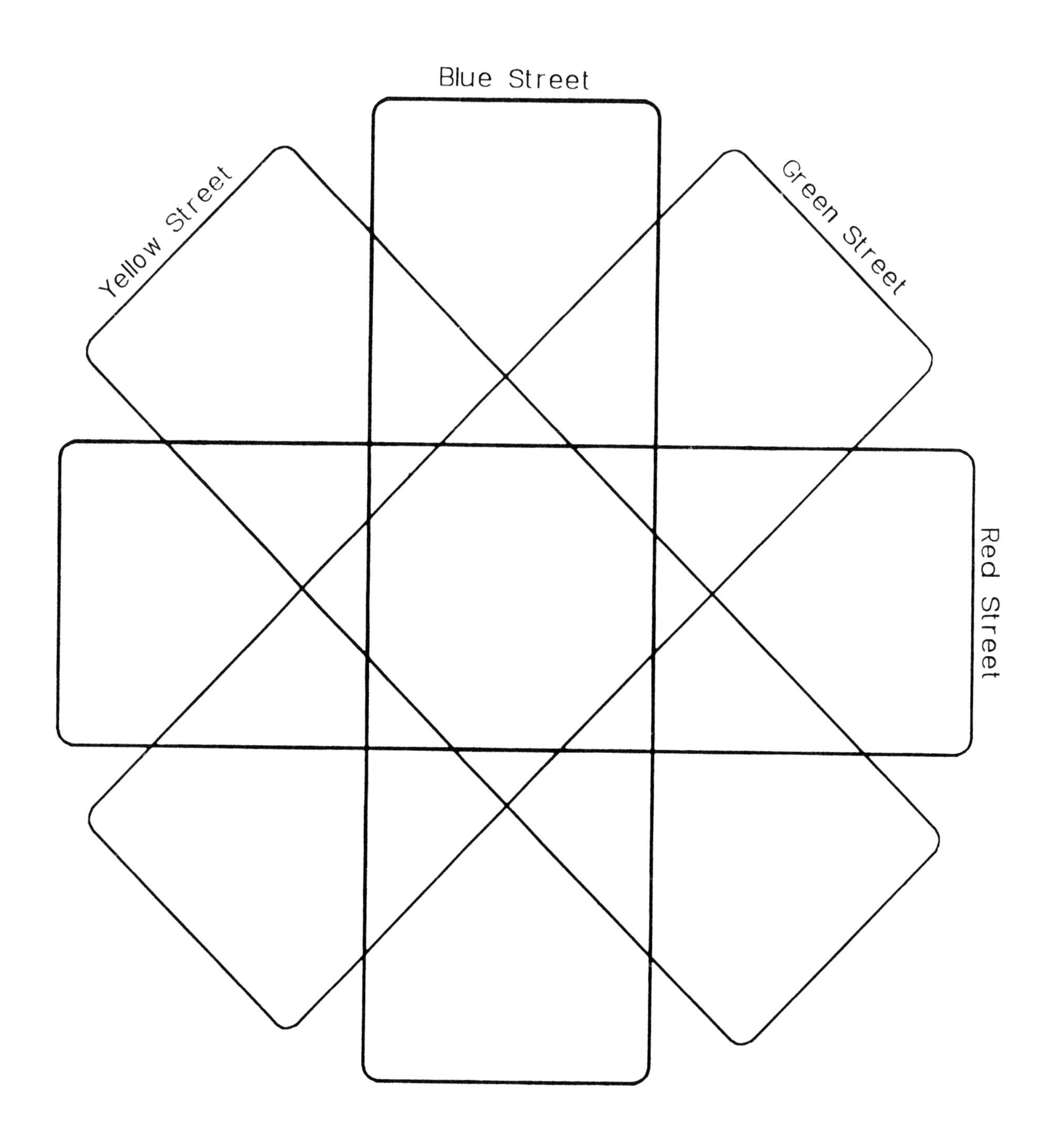

Cube Sheet 18

PARTNERS I

Cube Sheets 19 and 20

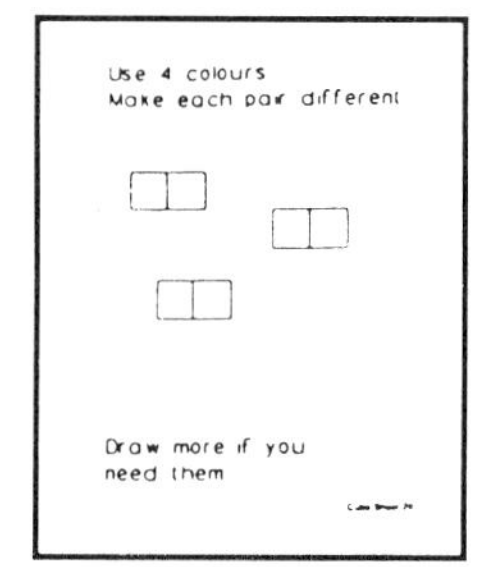

****

These activities encourage children to look for patterns.

****

- Pupils need 6 of each of the following colours:- red, yellow, green.
- Ask them to make as many different coloured pairs as they can using three colours. (Sheet 19).
- Extend the activity to use four or more colours. (Sheet 20). This will generate the triangular numbers

Children should not be given the pattern:-

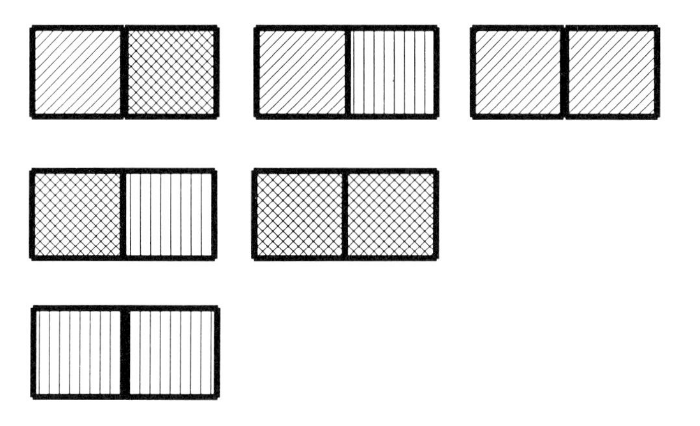

cont....

The value of this activity is in the search for a system. It does not matter that the children do not see a pattern on their first encounter with this type of problem.
Other activities produce the same pattern, which the children will eventually discover for themselves.

For example:-

1. How many different pairs of numbers can be made with two dice?

2. If there are four ice-cream flavours, how many different double cones can be made?

3. There are five children, how many handshakes will there be if everyone shakes hands with everyone else?

4. Make your own set of dominoes, numbers 1 - 4.

Use 3 colours
Make each pair different.

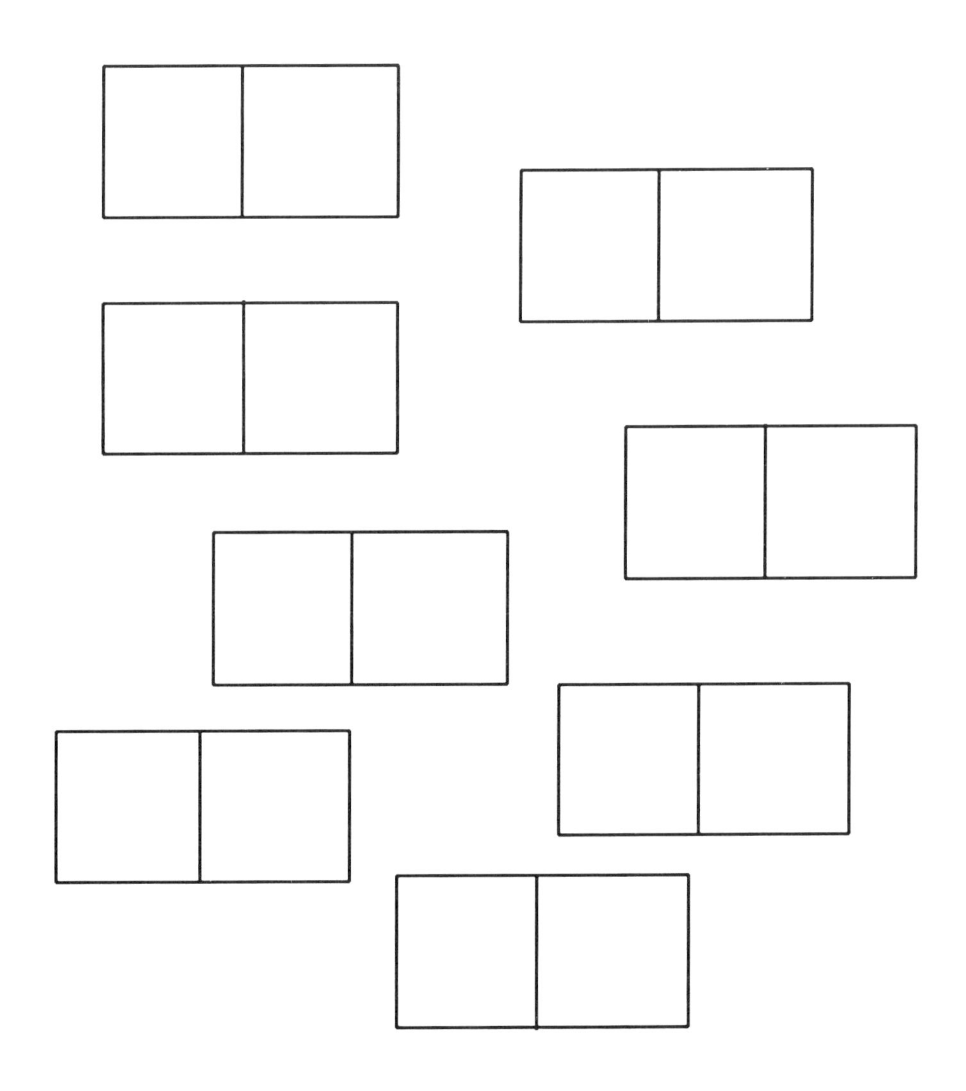

Are there too many pictures?

Cube Sheet 19

Use 4 colours
Make each pair different.

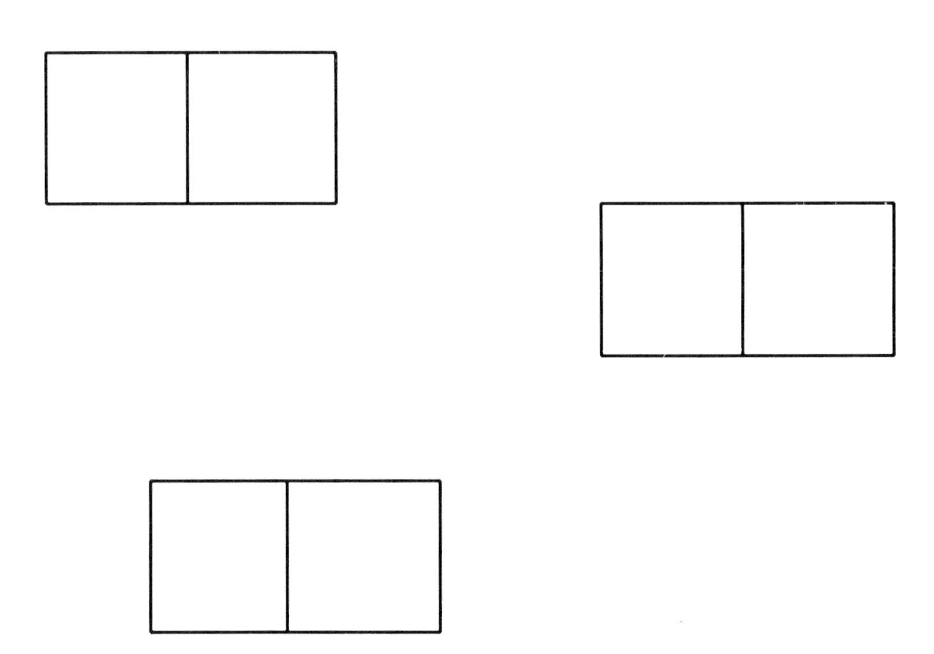

Draw more if you need them.

Cube Sheet 20

PARTNERS II

Cube Sheets 21 -26

Price List

Red	1p
Yellow	2p
Green	3p
Blue	4p
Pink	5p
Black	6p

Price List

Red

Yellow

Green

Blue

Pink

Black

Take your partners

Find three pairs worth 7p

Which pairs are worth more than 8p? Draw them

Which is the cheapest pair?

Find a pair worth the same as a blue

Draw 3 cubes worth 9p altogether

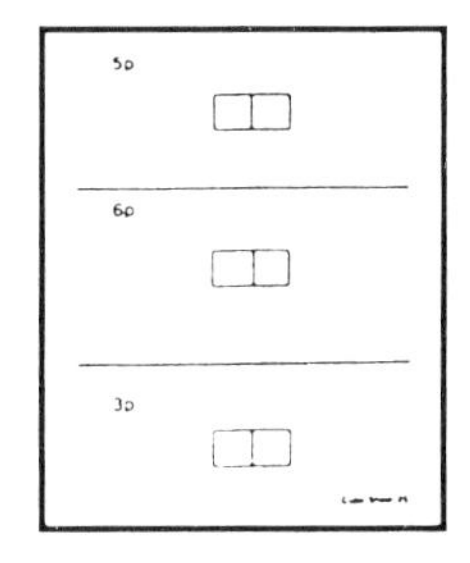

5p

6p

3p

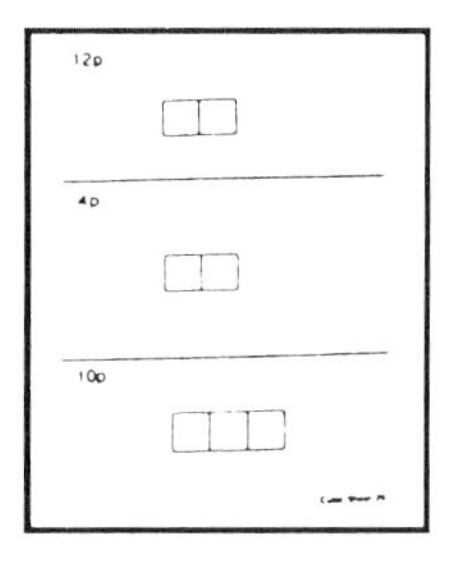

12p

4p

10p

****

These sheets are extensions to the activity Partners 1

****

- Use cube sheet 21, or make your own price list using cube sheet 22.
- The children find the value of each pair. This can be recorded on 2cm squared paper. eg.

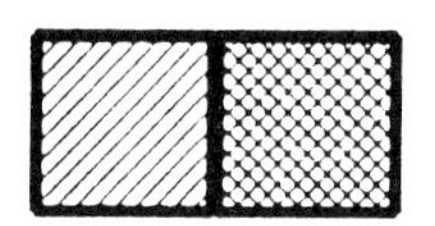

$1 + 2 = 3$

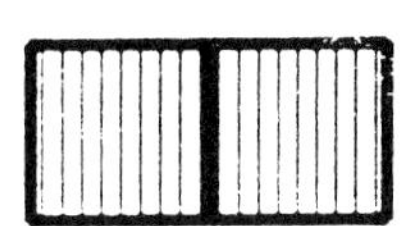

$3 + 3 = 6$

$2 + 2 = 4$

$2 + 3 = 5$

- Cube sheets 21 - 24 are designed to be made into a booklet, with problems for children to solve. Teachers may wish to add pages of their own design or to alter the values of the cubes to suit the needs of their children.

Price List

Red	1p
Yellow	2p
Green	3p
Blue	4p
Pink	5p
Black	6p

Cube Sheet 21

Price List

Red

Yellow

Green

Blue

Pink

Black

Cube Sheet 22

Take your partners

Find three pairs worth 7p

Which pairs are worth more than 8p? Draw them.

Cube Sheet 23

Which is the cheapest pair?

Find a pair worth the same as a blue.

Draw 3 cubes worth 9p altogether.

Cube Sheet 24

5p

6p

3p

Cube Sheet 25

12p

4p

10p

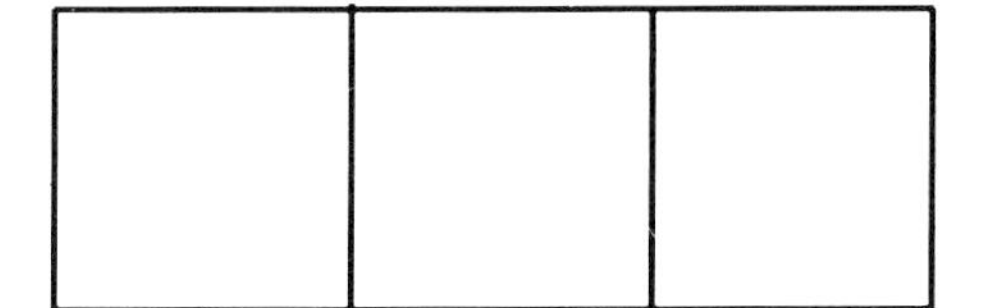

Cube Sheet 26

THE GROUPING GAME

Cube sheets 27-30

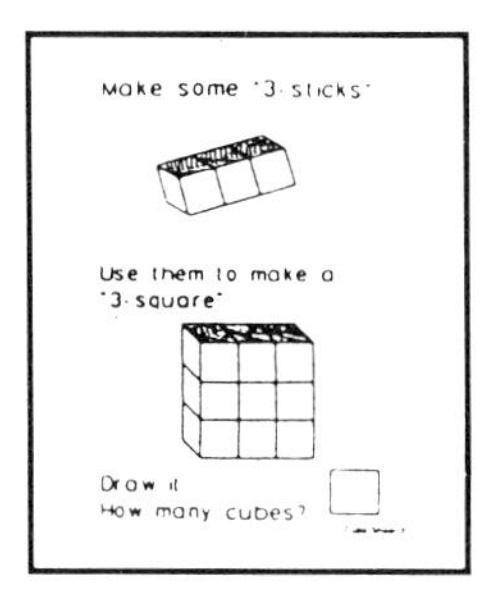

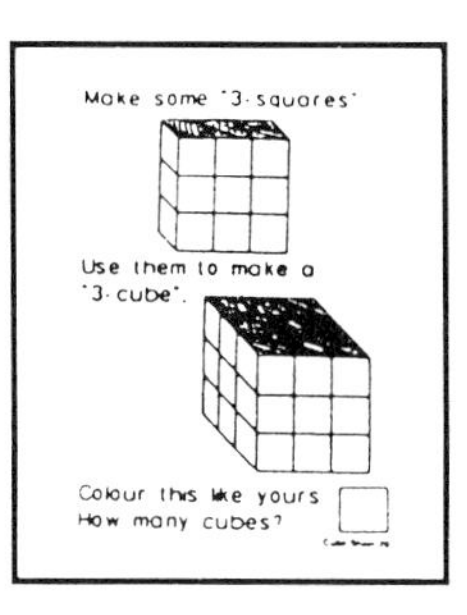

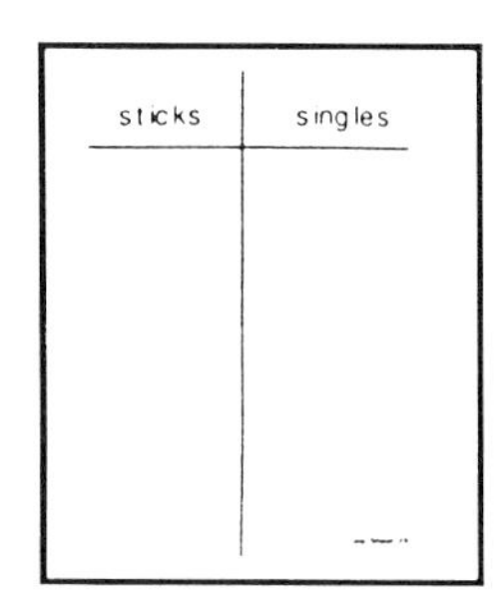

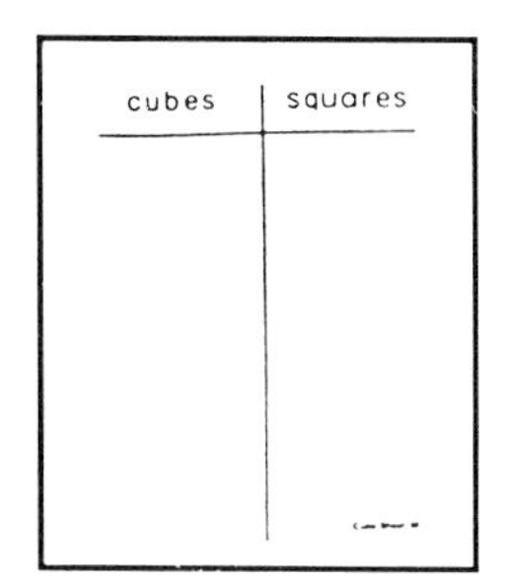

****

These activities are an introduction to Place Value

****

Encourage the children to make sticks, squares and cubes of different sizes. For example,

a "5 - stick"

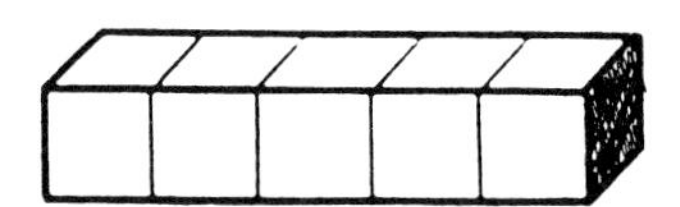

a "2 - square"

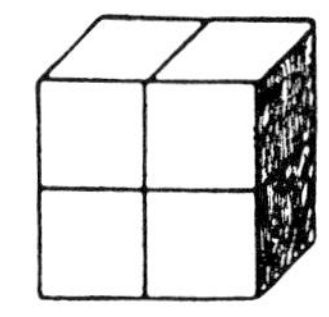

a "3 - cube"

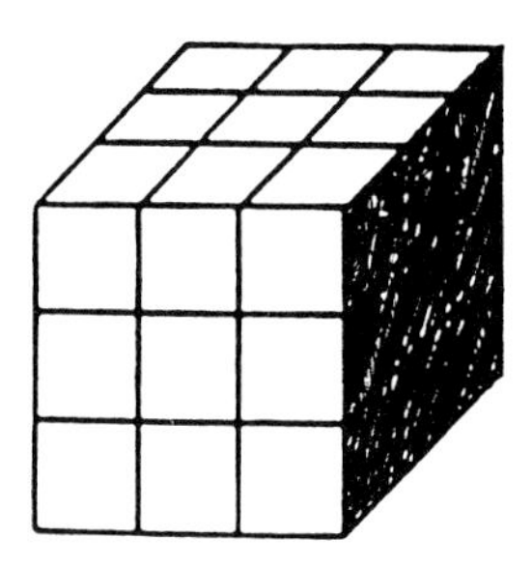

cont....

The cube sheets 29 and 30 should be made into game boards:-

cubes	squares	sticks	singles

To Play the game of 3

- This is a game for 4 - 6 players.
- One player (or the teacher) is banker, and has the cubes (500).
- Other players have a grouping board each.
- One die is needed.

1. Players take it in turns to throw the die and ask the banker for the appropriate number of cubes.
2. These are placed in the singles column.
3. When there are three or more singles in this column, join three to make a stick and place it in the Stick column.
4. If there are three or more sticks, join three to make a square. This is placed in the Squares column.
5. Play continues until one player has three squares, which are made into a cube and placed in the cube column.

cont....

The following rules are recommended:-

1. Players wait until the previous player has finished organising their cubes.
2. Bankers must check that everything is in the right place before the die is thrown.
3. When the game is understood, anything placed or left in the wrong column is taken away by the banker.

Extend the game by reversing it.

Each player starts with a "3 - cube" and gives to the banker the number of singles indicated by the die. To do this the "3 - cube" must be broken down into squares, sticks and singles, and put in the right place. The first player to get rid of all their single cubes wins the game.

Make some "3-sticks".

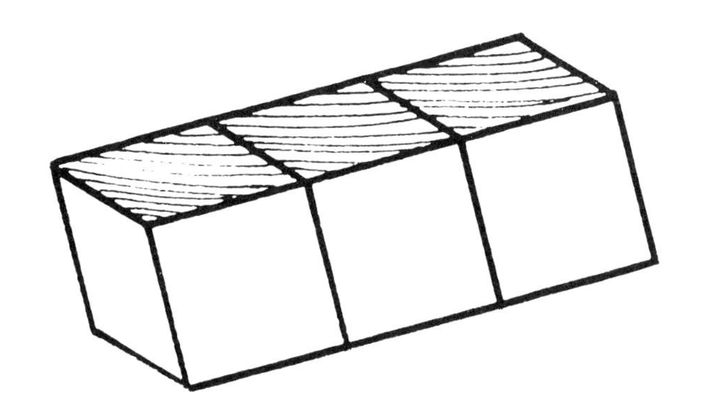

Use them to make a "3-square".

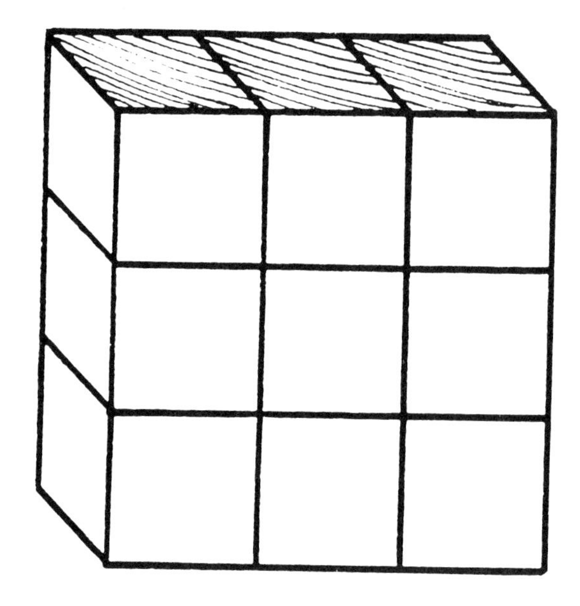

Draw it.
How many cubes?

Cube Sheet 27

Make some "3-squares".

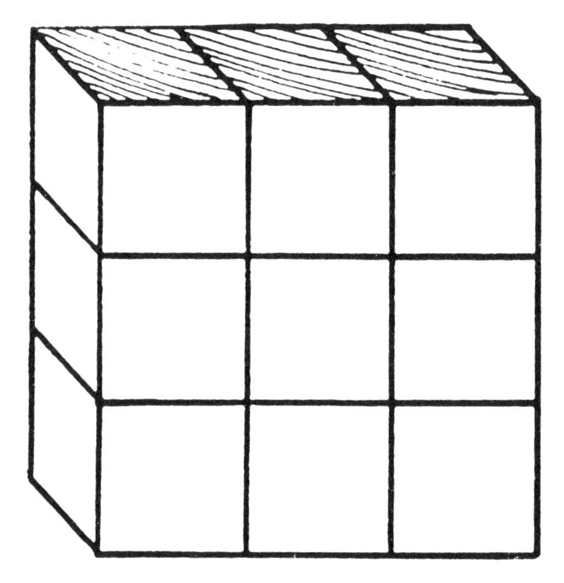

Use them to make a "3-cube".

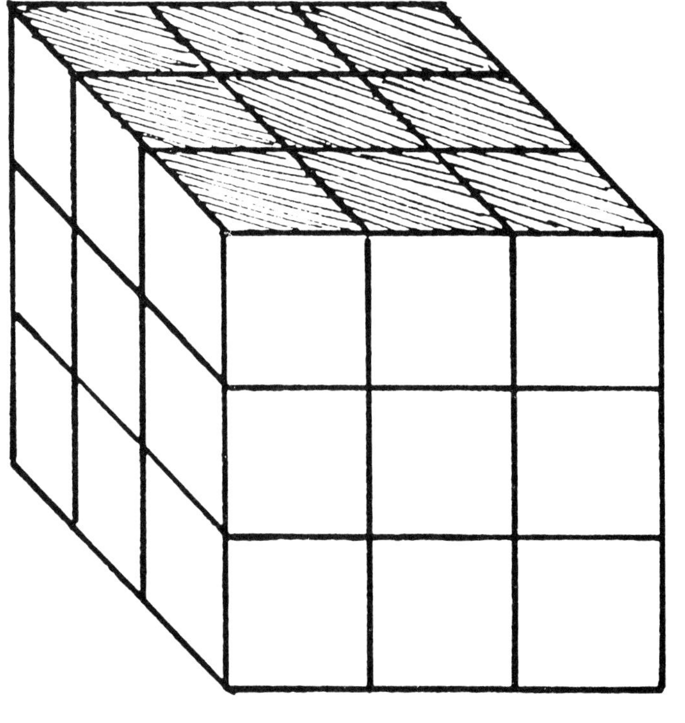

Colour this like yours.
How many cubes?

Cube Sheet 28

sticks	singles

Cube Sheet 29

cubes	squares

Cube Sheet 30

LEAVES

In this section leaves are used to develop ideas of estimation, counting, measurement and number.
Most of the activities are just as effective with other shapes, for example, foot or hand prints.

GUESS AND COUNT

Leaf sheets 1, 2 and 3

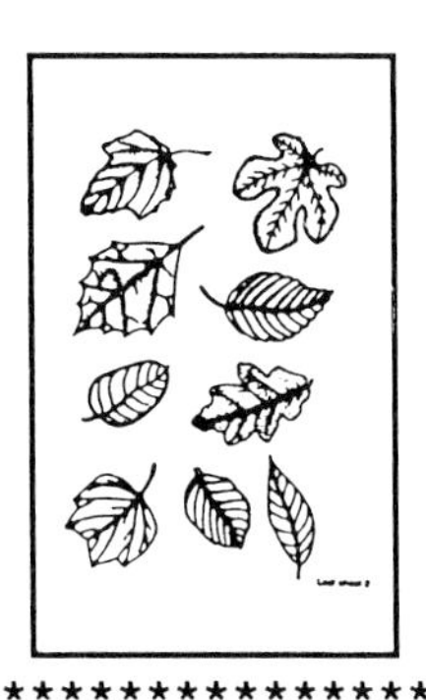

****

Activities on these sheets include:-

a.The invention and evaluation of a counting method

b.Discussion and refinement of that system

****

The following activities can be used with any of the leaf sheets.
Alter the number of leaves on a page if required.

Ask the pupils to :

1. Guess how many leaves are on the page.,

2. Find out through discussion how they arrived at their answers.

3. Let them count the leaves, the methods they use may include:-
   - pointing and counting
   - numbering
   - placing one counter on each leaf and then counting
   - grouping the leaves and counting the groups.

<u>EXTENSIONS</u>

1. Collect leaves. Estimate and count how many will fit on a blank page.

2. Cover a page with a specific number of leaves. For example, if the chosen number is 6, the children must choose 6 appropriately sized leaves to cover the page.

3. Ask children to draw a given number of leaves in such a way that they are easy to count, for example in pairs.

Leaf sheet 1

Leaf sheet 2

Leaf sheet 3

ESTIMATION AND COUNTING

Leaf Sheets 4, 5, 6 &7

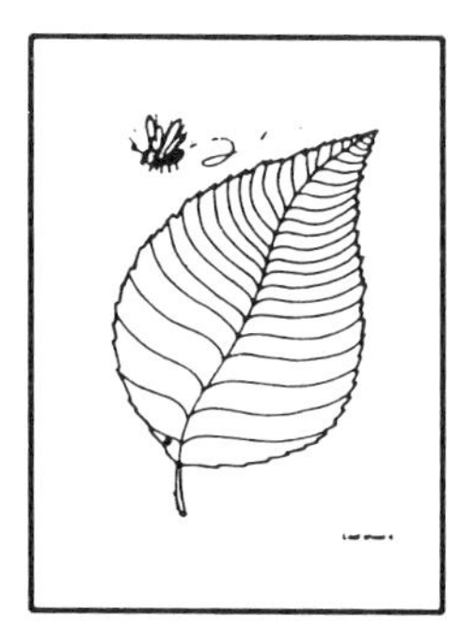

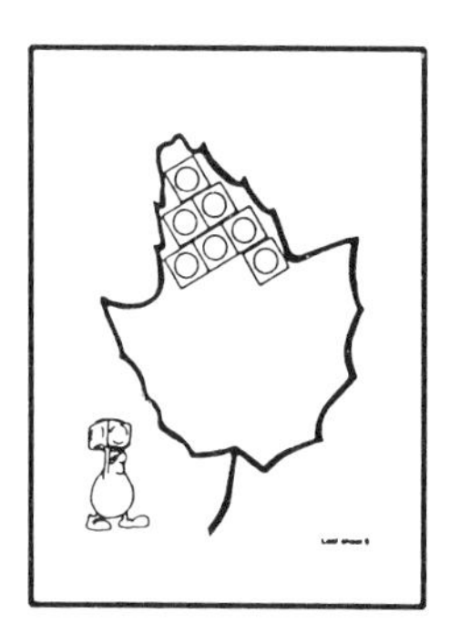

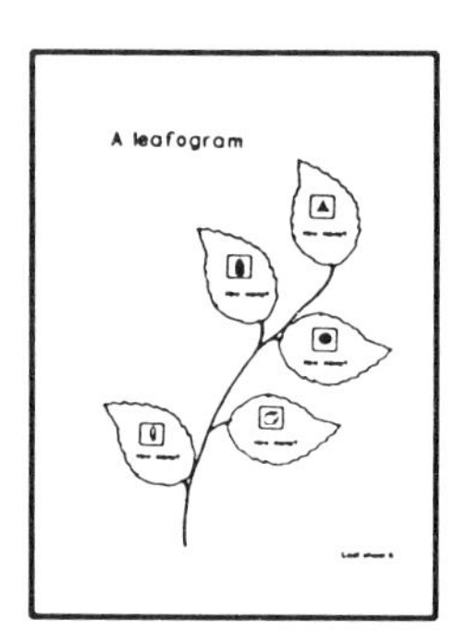

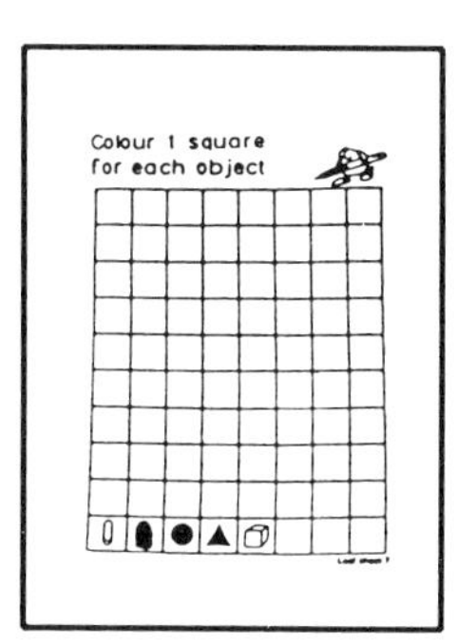

****

These sheets extend pupils' experience of estimating, counting, and covering spaces.

****

Needed: - One leaf per child (Use a real leaf or leaf sheets 4 or 5).
Coloured cubes.

Ask the pupils to: -

1. Estimate and record the number of cubes that will cover the leaf.

2. Put 6 cubes anywhere on the leaf and review their estimate.

3. Fill the leaf with cubes and consider the following questions....
   How good was your estimate?
   Did the 6 cubes make the estimate more accurate?
   Would it have been better to use 10 cubes?
   How would you find the number of cubes that cover the leaf if you only had 5 cubes?

<u>FURTHER PROBLEMS</u>

i) Repeat the activity with different leaves.
ii) How many thumb prints can you fit on the leaf?
iii) Try other objects, for example, paperclips, smaller cubes or counters. (Leaf sheets 6 & 7 can be used to represent this data)

The value of this activity is lost if children do not record and discuss their work.

iv) Find 20 things which cover the leaf completely.

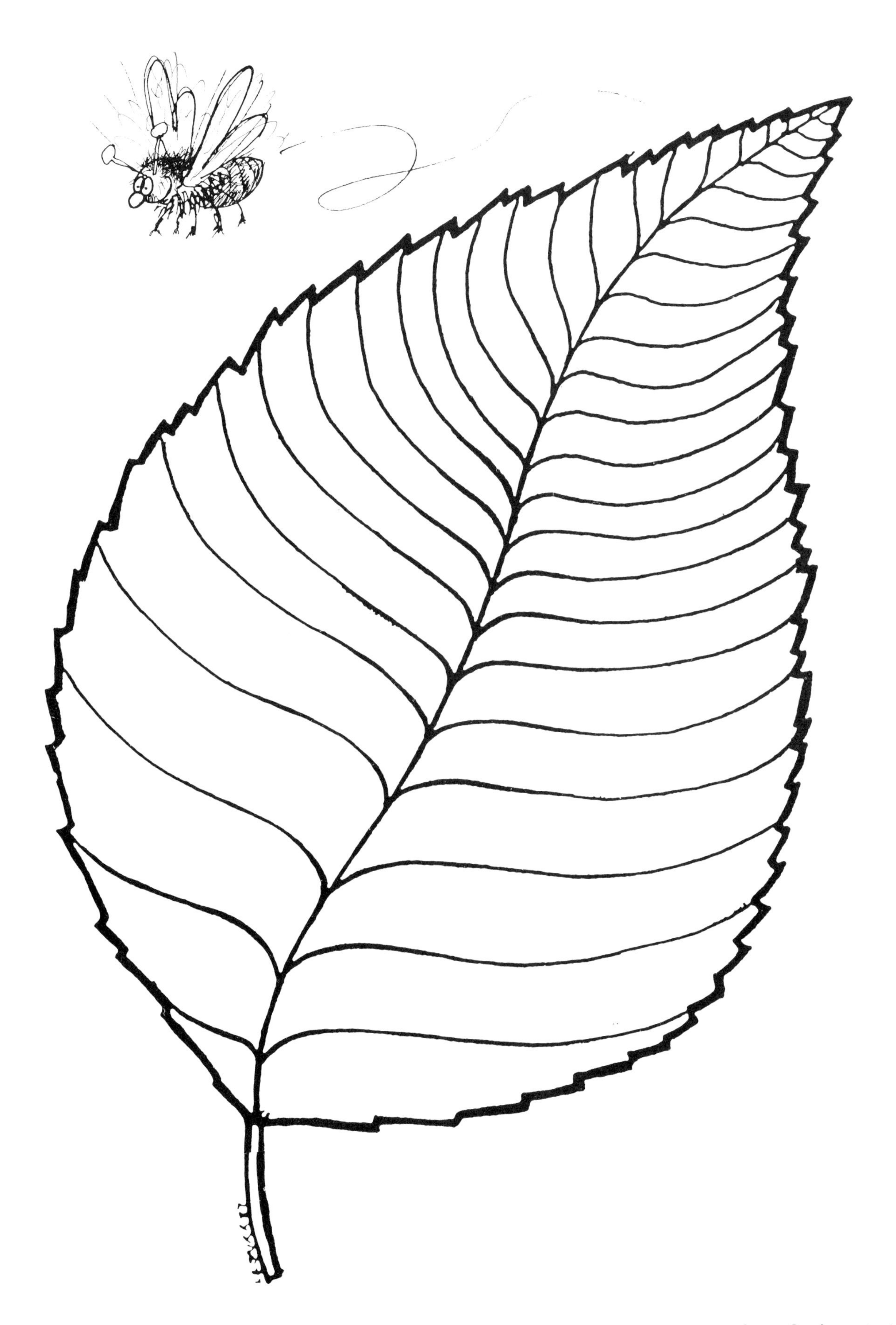

Leaf sheet 4

Leaf sheet 5

A leafogram

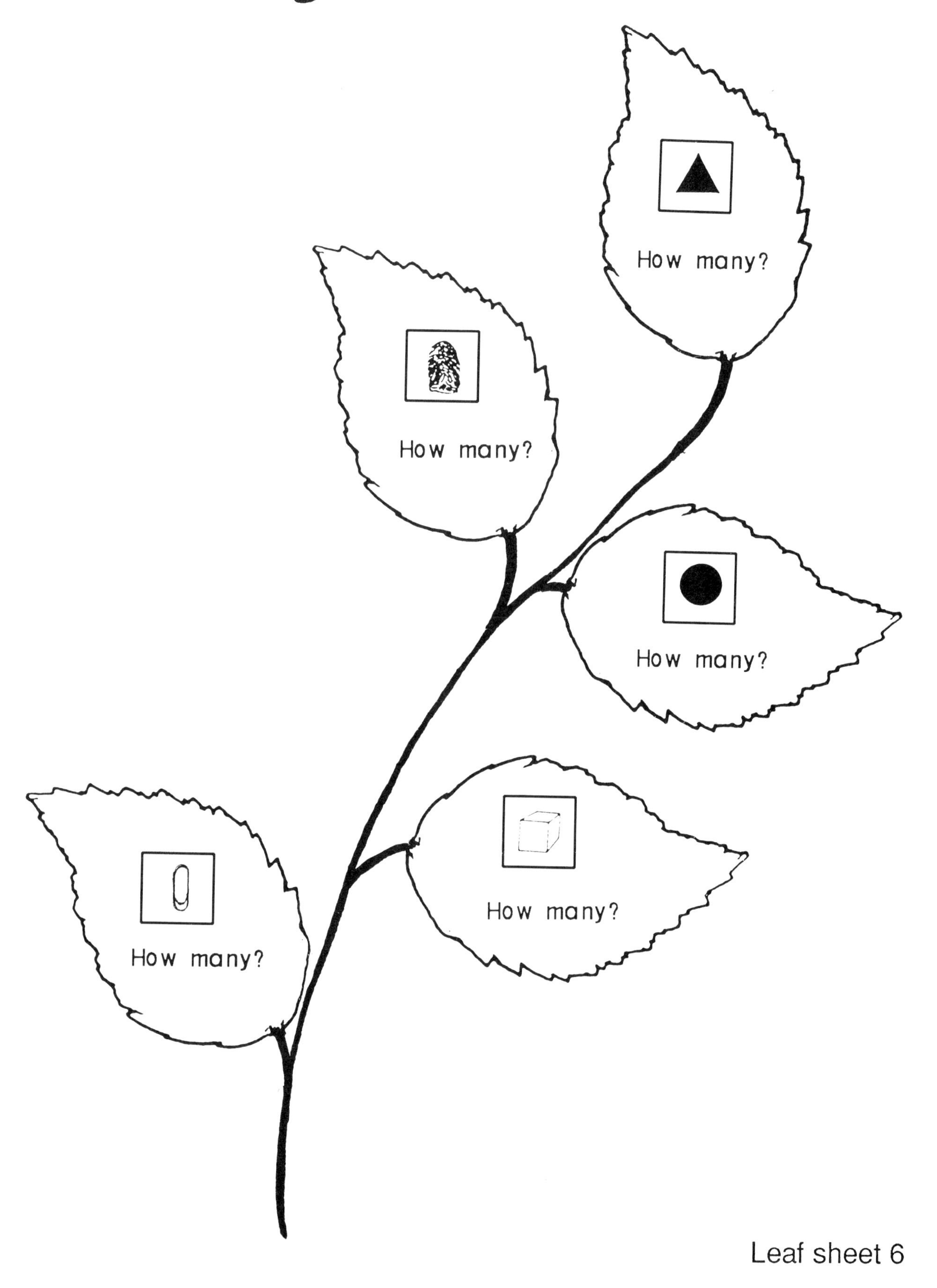

Leaf sheet 6

Colour 1 square for each object

Leaf sheet 7

PERIMETER

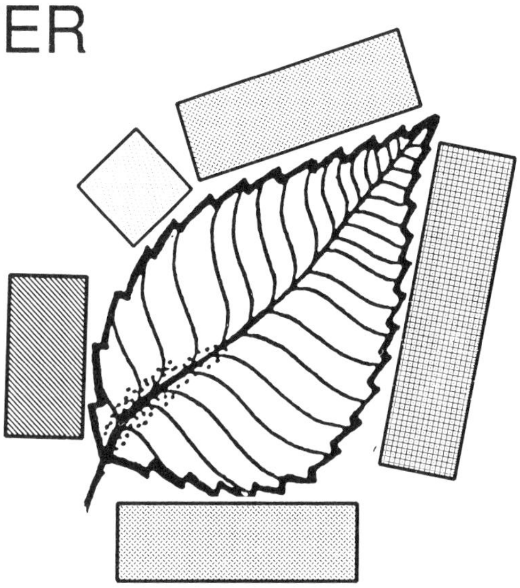

Needed: Real leaves
or
Leaves cut from Leaf Sheet 2
Rods of various lengths

Ask the children to:-

1. Put some rods around the leaf.
2. Imagine the rods are made into a 'train'.

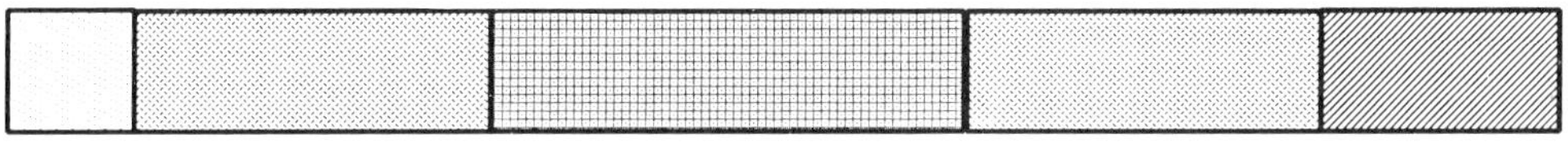

Estimate and record how long that train would be.
Estimates can be recorded like this:

Name........................	
Estimate	
Rods	

3. The values of the rods can be added together to find the perimeter. This will be in centimeters. Whether or not it is appropriate to use this fact is left to the teacher.

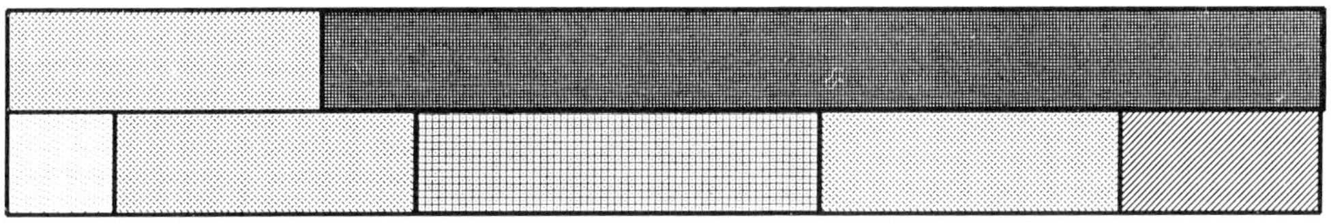

Place value is reinforced if the orange rods are used.

4. Discuss the results: How good were the guesses?
   Is there a better way to find the perimeter?
   Why is it better?
   Could we find 4 leaves with the same perimeter?

AREA

Leaf Sheet 8

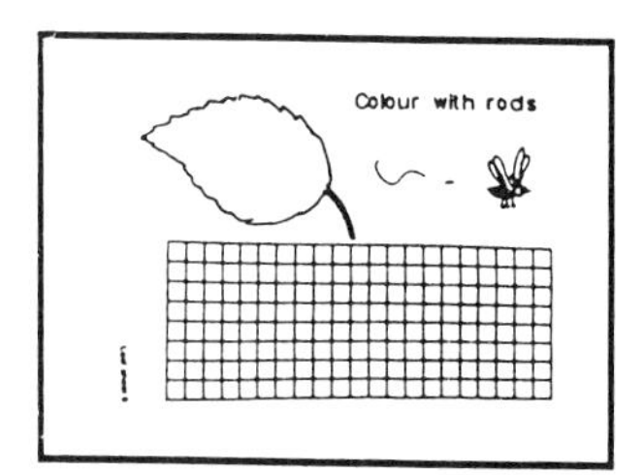

Needed: Cuisenaire rods

Ask the children to make the leaf on Leaf sheet 8 attractive, by filling it with rods.

Ask them:-

1. To imagine that the rods are used to make a train.
   Is there something in the classroom that might be the same length.

2. Estimate and record the length of that train.
   Their methods of recording may include:-

a. marking the beginning and the end of the train,

b. drawing and colouring the train,

c. counting the numerical value of the rods,

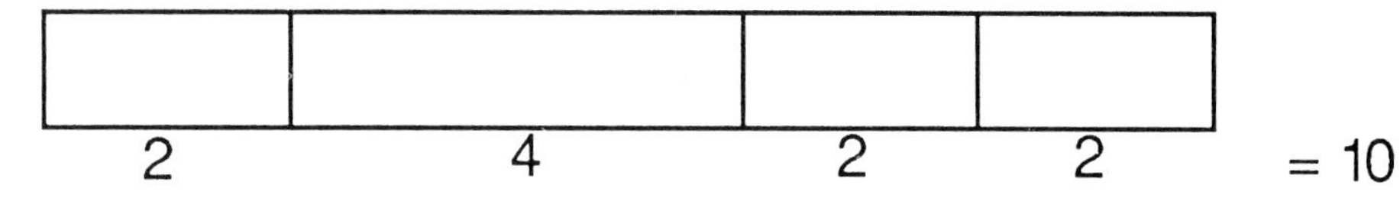

3. Cover the leaf in a different way.
   Estimate and record the length of the new train.

4. Discuss the results.

<u>Further Problems</u>

1. Use as many colours as possible.
2. Use as few colours as possible.
3. Fill it in 5 different ways.
4. Use a specific number of rods eg. 10

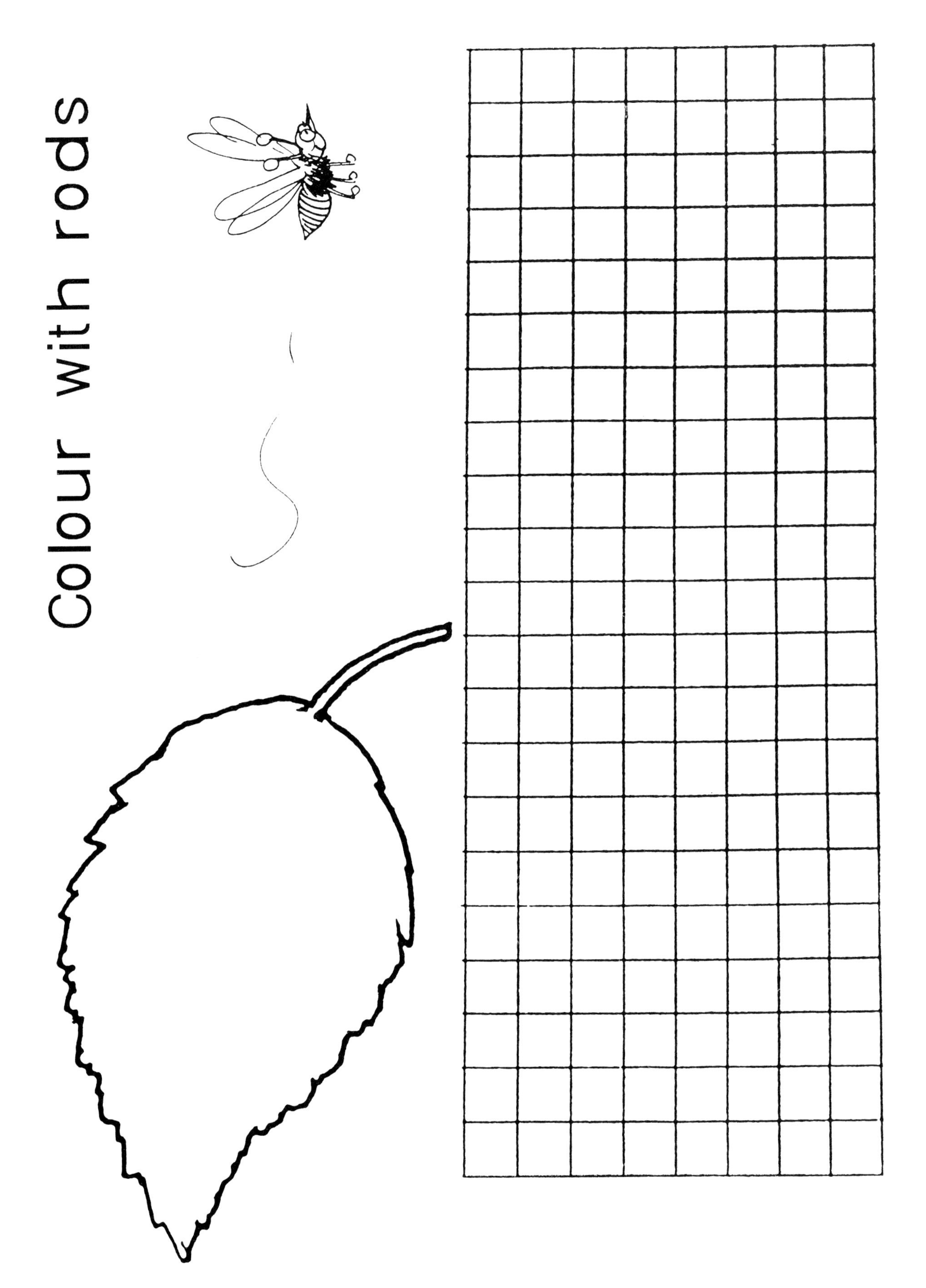

Leaf sheet 8

FILLING SPACE

Leaf sheet 9 or 10

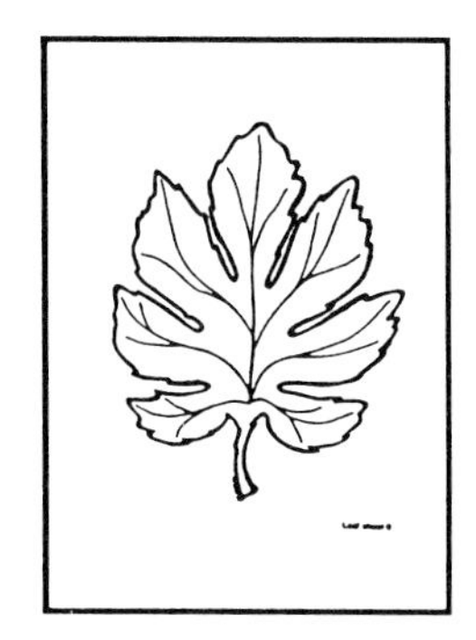

****

As well as providing more activities for filling space, this activity helps develop an understanding of area.

****

Needed: Triangles
or
Triangle grid paper

(*Transparent grids can be made by copying one of the grid sheets at the back of the book, onto acetate sheets using a heat or photo copier.)*

1. Estimate and record the number of triangles the leaf covers.

2. Count the triangles.

3. Ask the children to invent ways of counting that make checking easy, which may include:-

- numbering the triangles

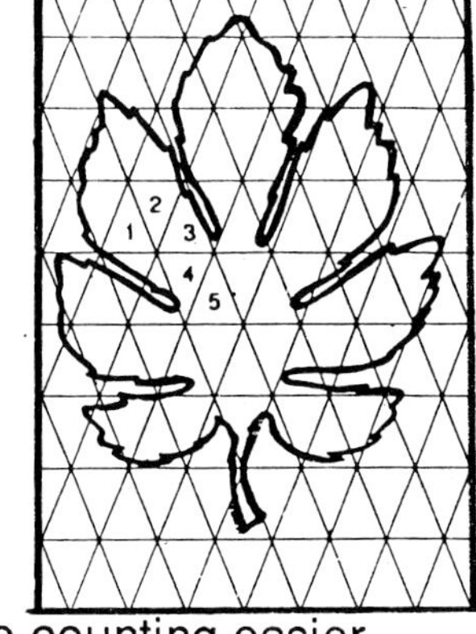

- colouring them in groups to make counting easier.
  For example:

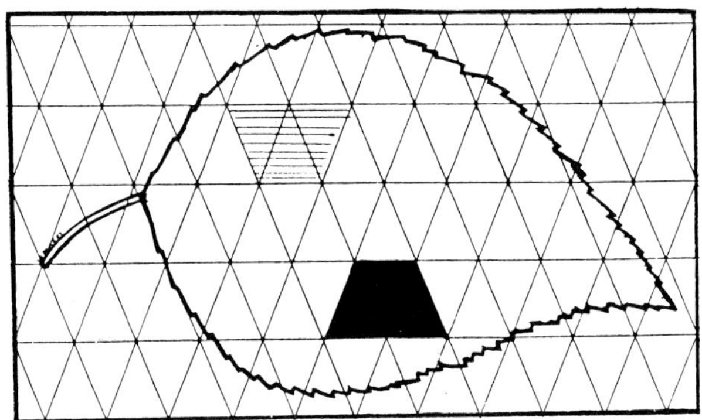

The activity can be repeated using other grids and other leaves.

Leaf sheet 9

Leaf sheet 10

MY LEAF

Leaf sheets 11 to 17 are designed to be made into a booklet:-

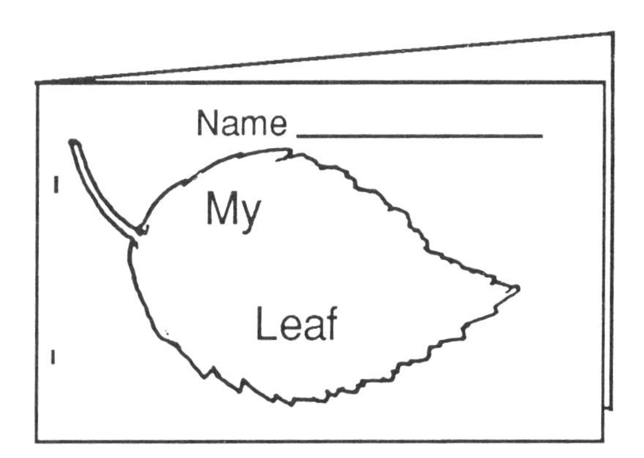

Children need one leaf each.

Alternative covers are given so that the same activities may be done using foot or handprints. For example:-

- Can you fill your footprint with money to make it expensive?
- Can you make it cheap?
- Measure your hands with Cuisenaire rods.

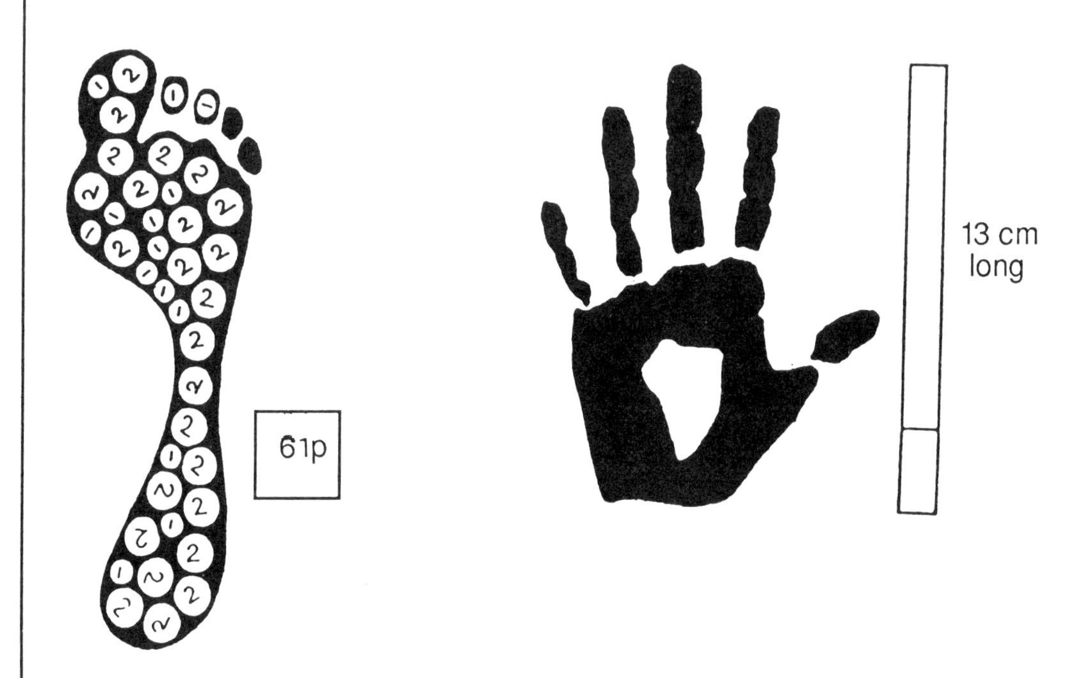

Teachers will probably think of their own questions to add to the books.

It is longer than

____

Draw it

Leaf sheet 11

It is shorter than

____

Draw it

I measured it
It was ________________

Draw it

Leaf sheet 12

Does it float? __________

Draw it

Does it float with cubes on it? __________

How many? __________

Draw it

Leaf sheet 13

How much money will fit on it?

____

Draw it.

Drop it.
What number can you count to before it reaches the ground?

Leaf sheet 14

What colour is it?

____

Draw it.

How many cubes fit on it? __________

Draw it.

Leaf sheet 15

What shape is it?

____

Draw it.

Draw something else that is special about your leaf.

Leaf sheet 16

Name ______________________________

My

foot

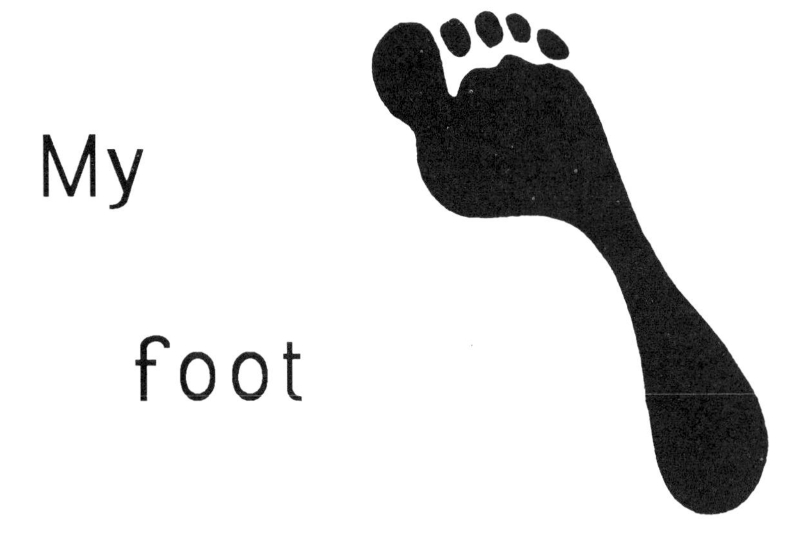

Name ______________________________

My

hand

Leaf sheet 17

MEASURING WITH LEAVES

Leaf Sheets 18 - 20

These sheets are designed to be made into a measurement book:-

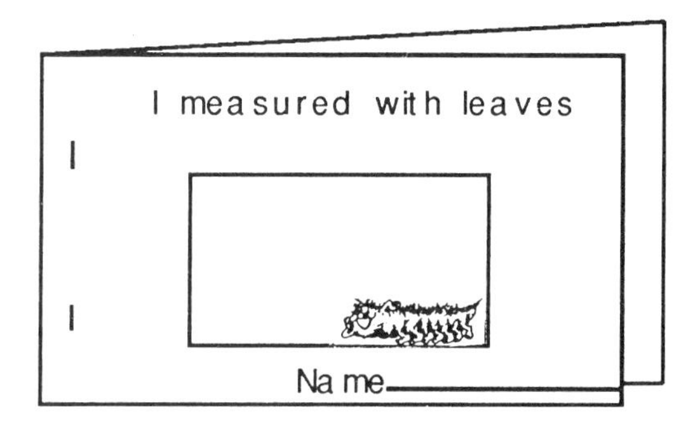

Teachers can add some pages of their own design.

Children can use single leaves or they can be stuck onto strips of card to make branches.

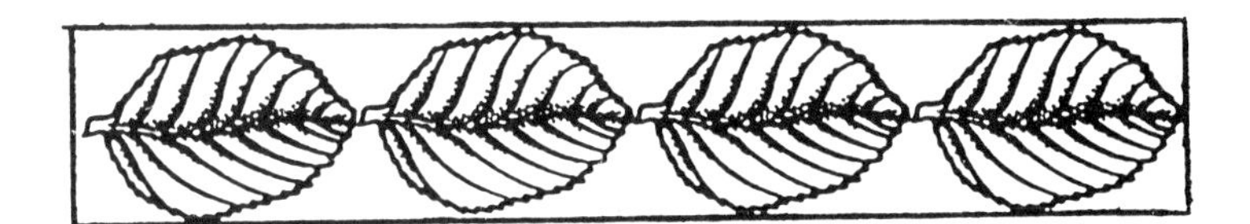

Either use a fixed length of card, children must find leaves to fill it, or cut a length of card to fit a given number of leaves

This ruler can be used to measure.

e.g. The rabbit hutch measures
2 branches and 3 leaves.

I measured with leaves

Name ____________

The table measured ____________

Leaf sheet 18

The doorway

measured ______

The shelf

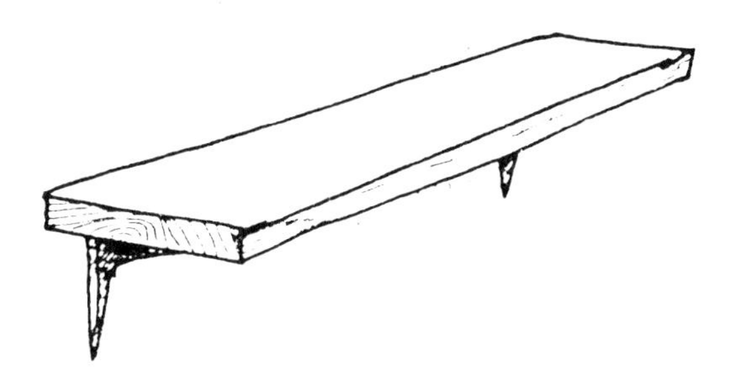

measured ______

Leaf sheet 19

Which used most? ____
Draw it.

Choose something to measure.
Draw it.

Measure it ____________

Leaf sheet 20

SPRINGTIME
Leaf sheets 21-24

A Place Value game for 4 players and a 'banker'.

You need: 4 tree trunks
20 branches
40 twigs
40 leaves
dice

The banker looks after the leaves, twigs, branches and trunks.

To play with the rule of 3

- Players take it in turns to throw the die and collect the number of leaves shown.
- When a player has collected 3 leaves, they can be exchanged for a twig.
- When 3 twigs have been collected, they can be exchanged for a branch.
- The winner is the first player to collect 3 branches which can be placed on a trunk to make a tree.

Variations

1. Play the game of 4 (Leaf sheet 24). Four leaves must be collected before exchanging for a twig etc.
2. Use 2 dice
3. Use different dice eg. one with only 1, 2 or 3 dots.
4. Play in reverse and call the game Autumn. Each player begins with a complete tree and removes leaves according to what is thrown on the die.

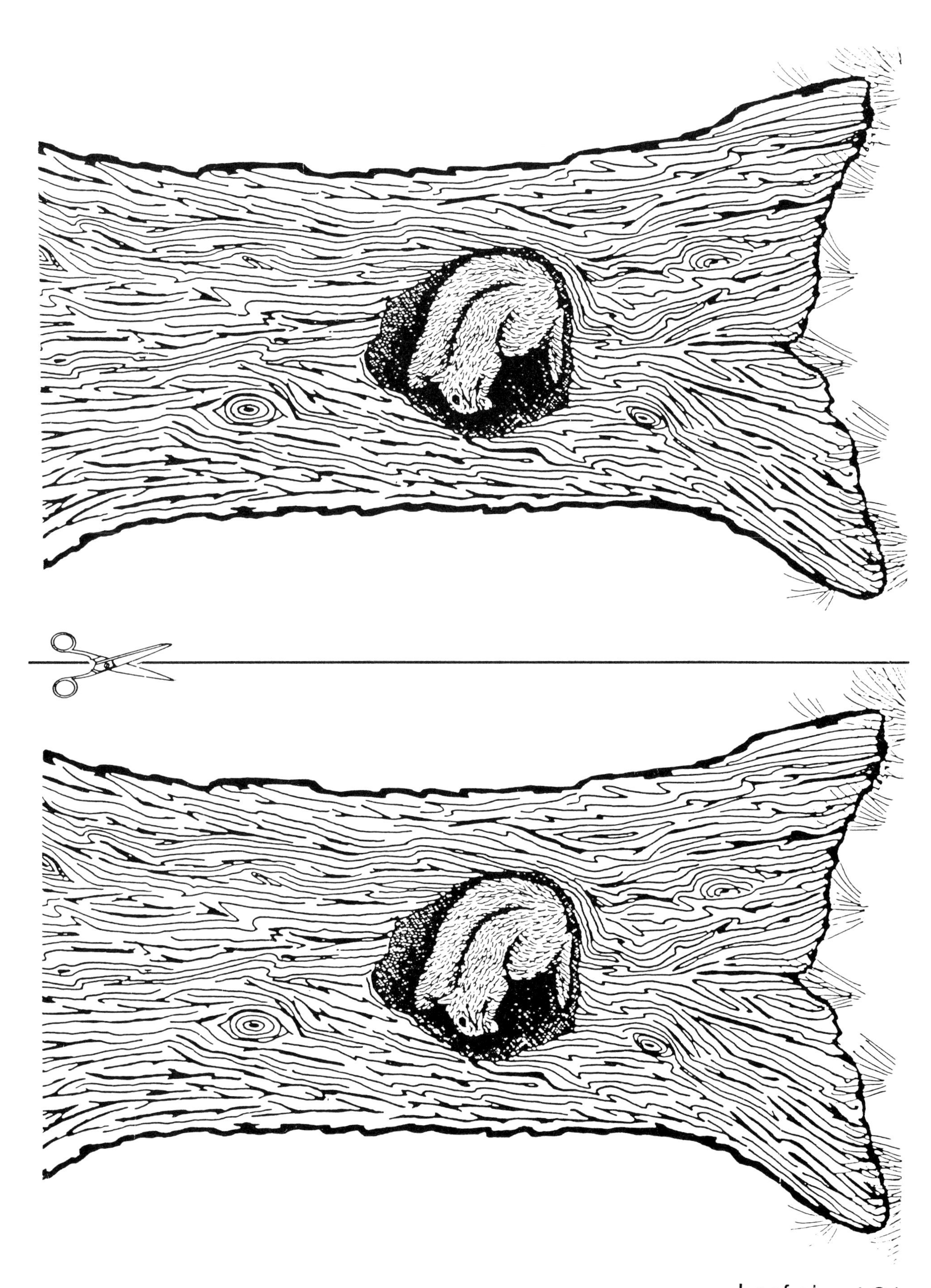

Leaf sheet 21

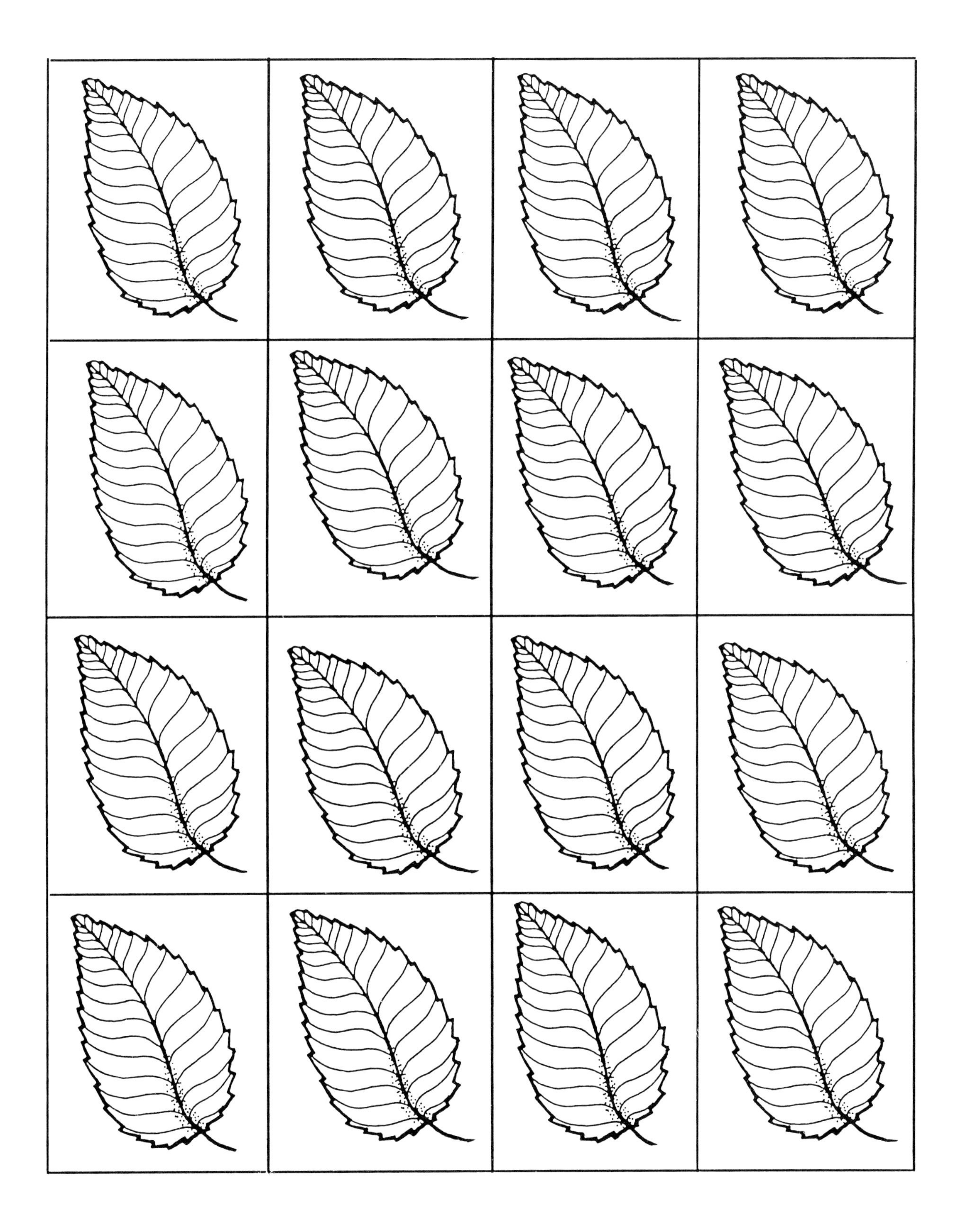

Leaf sheet 22

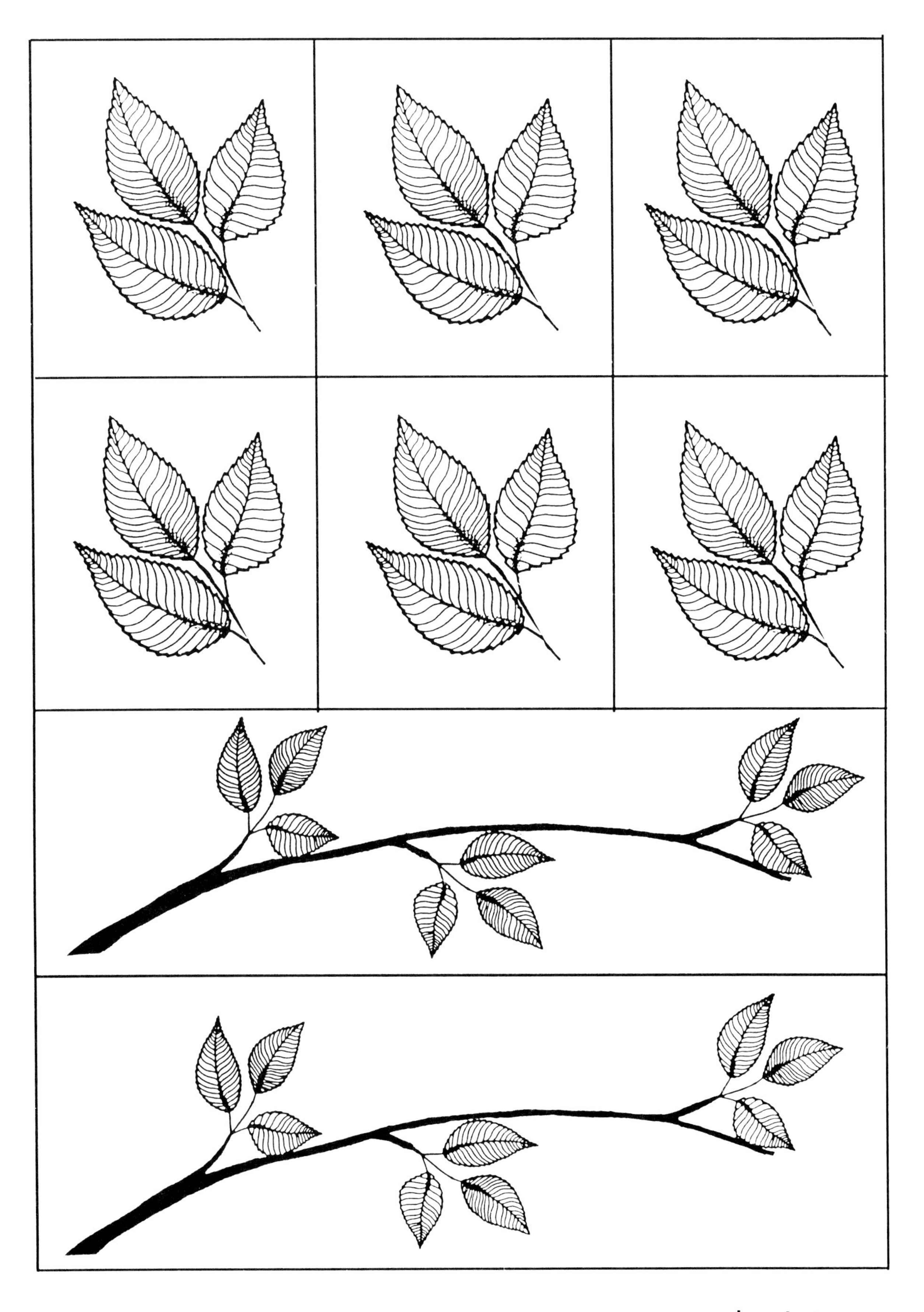

Leaf sheet 23

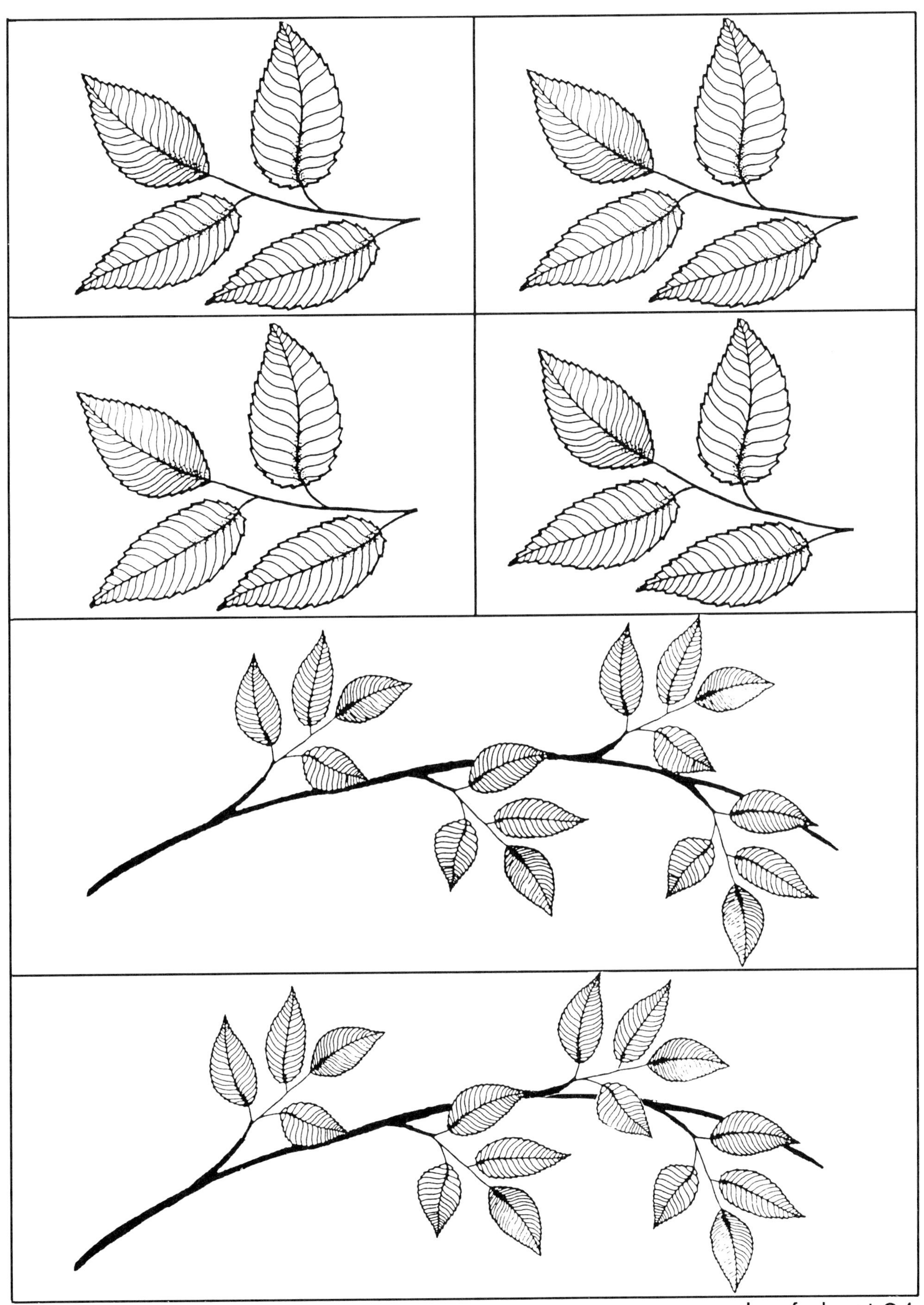

Leaf sheet 24

CUISENAIRE®

The ideas in this section are for use with Cuisenaire® Rods. Much of this curriculum can be used, or adapted for use with other equipment.

If you are unfamiliar with the rods, turn to the end of this section where there is a collection of introductory ideas.

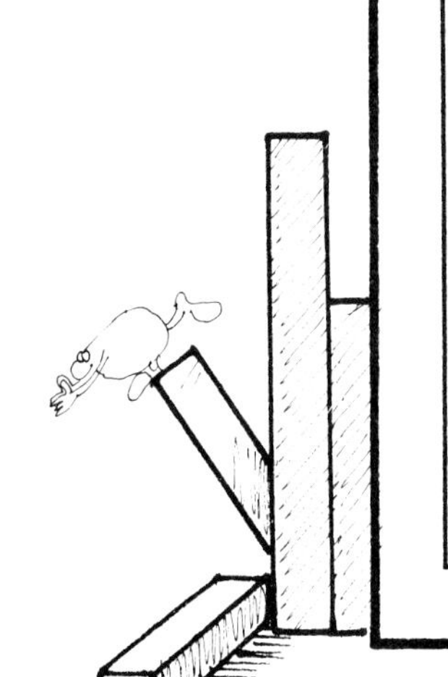

STORIES, POEMS AND CUISENAIRE® RODS

THE THREE BEARS

Cuisenaire® Sheets 1, 2, 3, 4, 5.

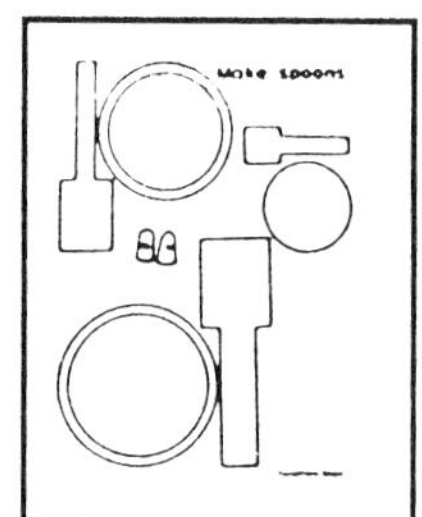
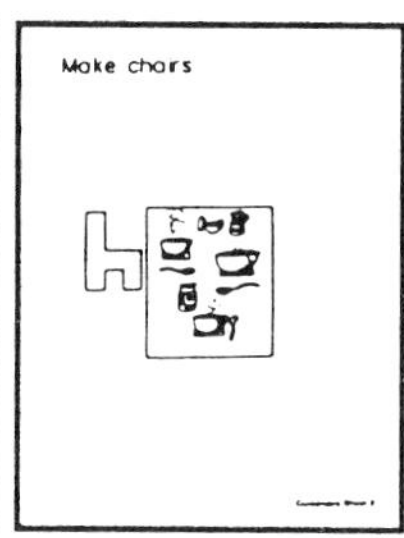
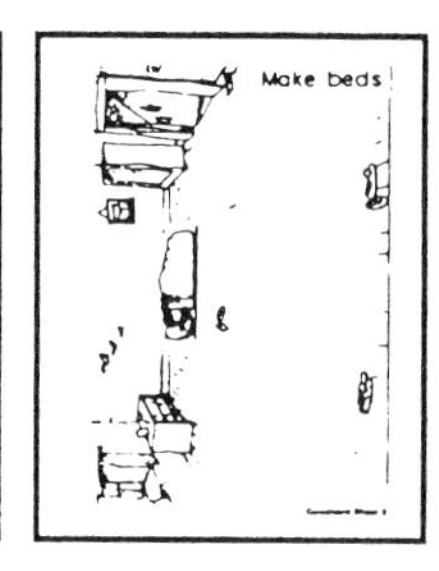

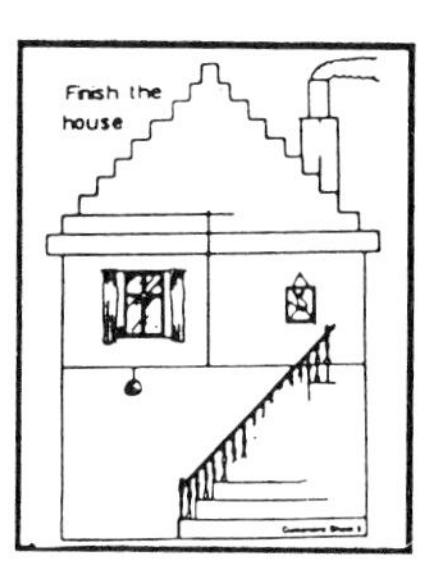

****

'The Three Bears' has been chosen as an example of how stories, rhymes and poems yeild a wealth of mathematical activities. We suggest that the children have enjoyed the story at least once, before these mathematical activities are introduced.

****

i. Before telling the story, the children choose a rod for each of the bears and hide them behind their backs.
Each time the name of one of the bears is mentioned in the story, the children bring that "bear" from behind their back without looking.

ii. Worksheets 1, 2 and 3, show scenes from the story.
The children complete the pictures using rods to fill the spaces.

iii. Worksheet 4. Cut out the bears and ask the children to make furniture for the bears to use. For example, make a chair for Daddy bear.

iv. Worksheet 5. Use the rods to furnish the house, tile the roof and carpet the stairs.

v. If the children are familiar with the numerical value of the rods, the following questions can be asked:-

- What is the value of each chair?
- Using rods worth 100, make chairs (beds, furniture) for everyone.

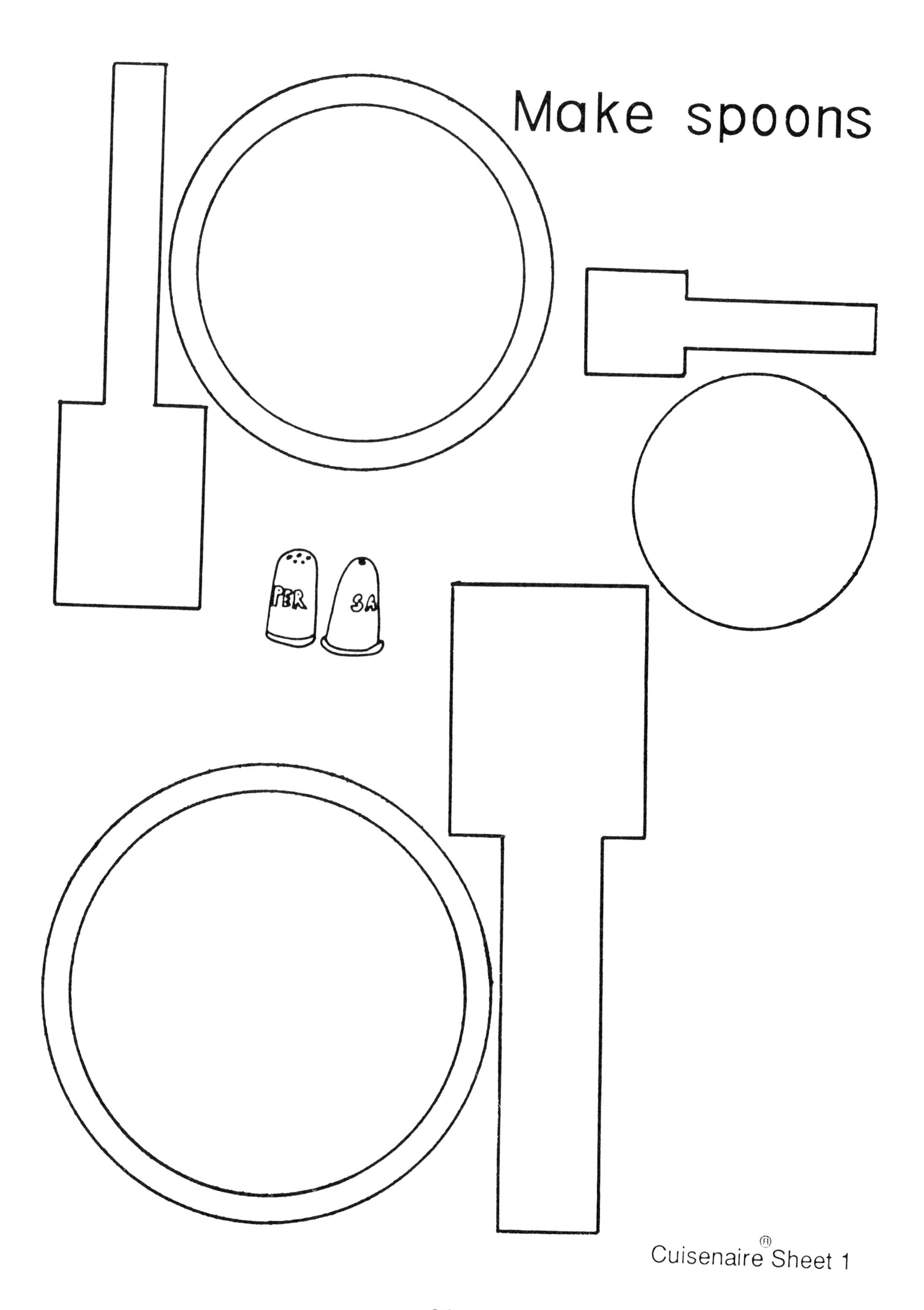
Make spoons
PER
SA
Cuisenaire® Sheet 1

Make chairs

Cuisenaire® Sheet 2

Make beds

Cuisenaire® Sheet 3

Cuisenaire® Sheet 4

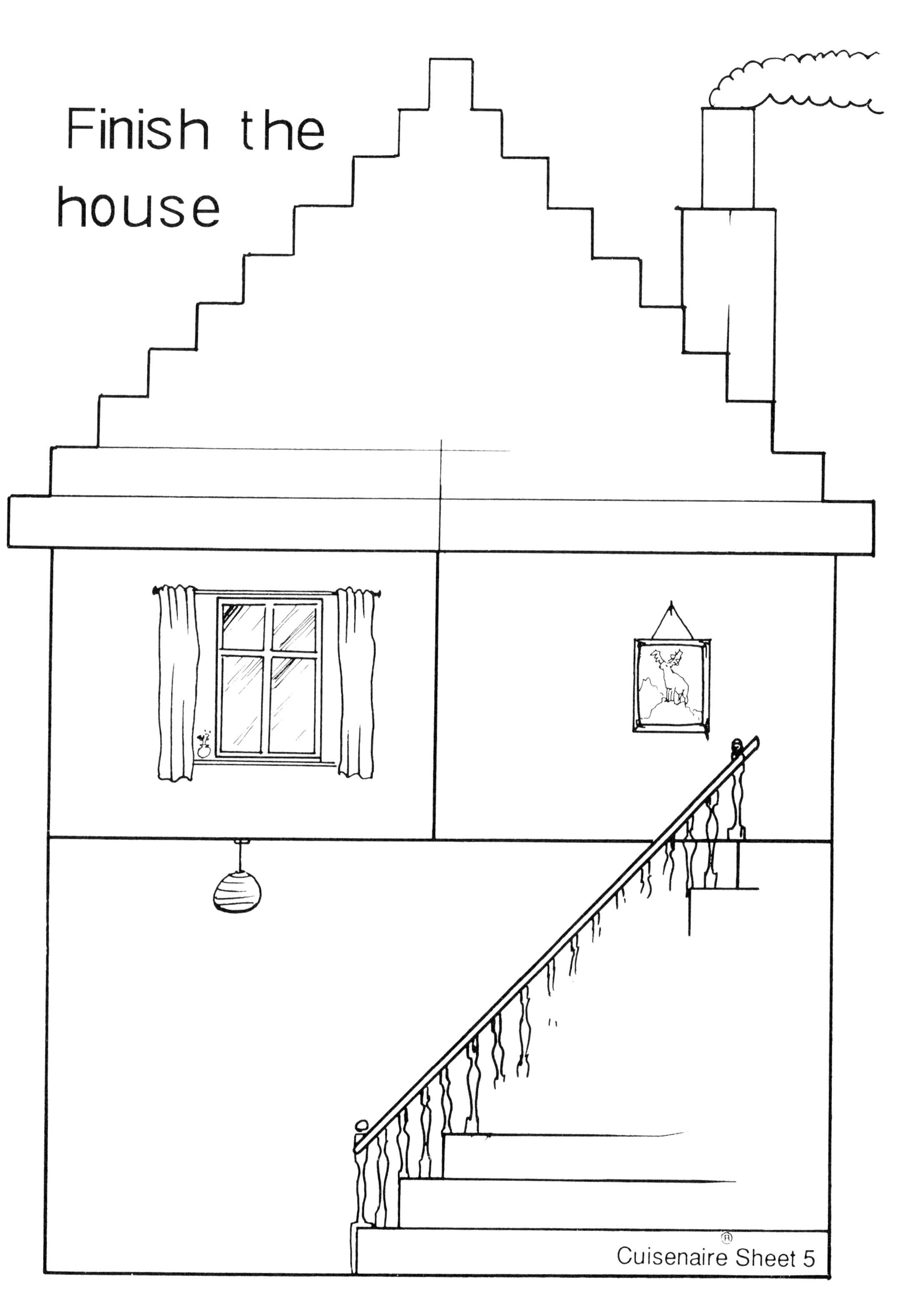
Finish the house
Cuisenaire Sheet 5

GUESS IT!

A game for 2 - 6 players

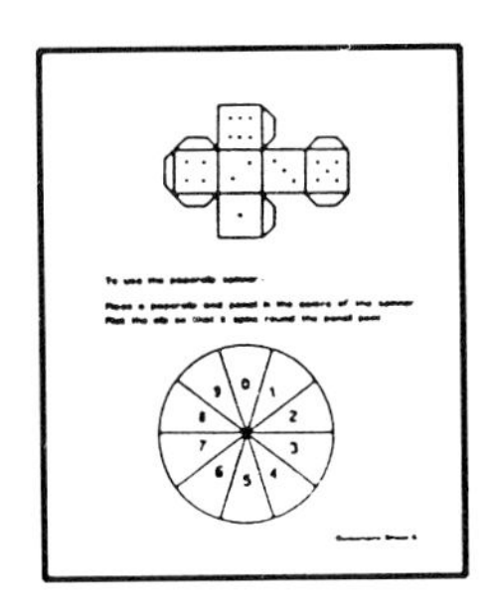

Use Cuisenaire® sheet 6.

Make a die or spinner using the nets given.

****

This activity familiarises children with the rods.

****

To play:-

Players take it in turns to:-

1. Spin or throw a number.
2. Take that number of white rods and place them in a line...

3. Guess which colour rod will be the same length as the line, and then check.
4. If the guess is correct, the player keeps the coloured rod. If not, the rod must go back in the box.
5. After 5 rounds, the player with the most coloured rods or the longest train, wins the game.

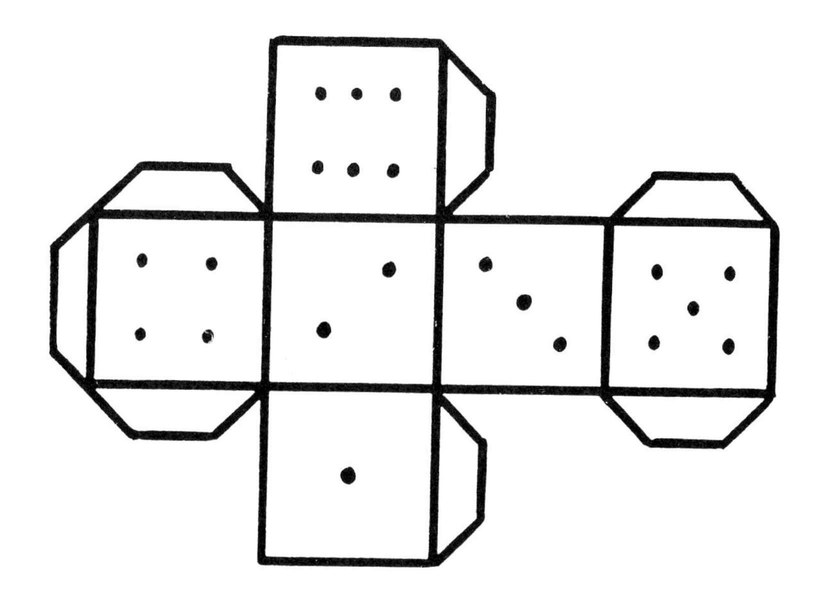

To use the paperclip spinner:-

Place a paperclip and pencil in the centre of the spinner. Flick the clip so that it spins round the pencil point.

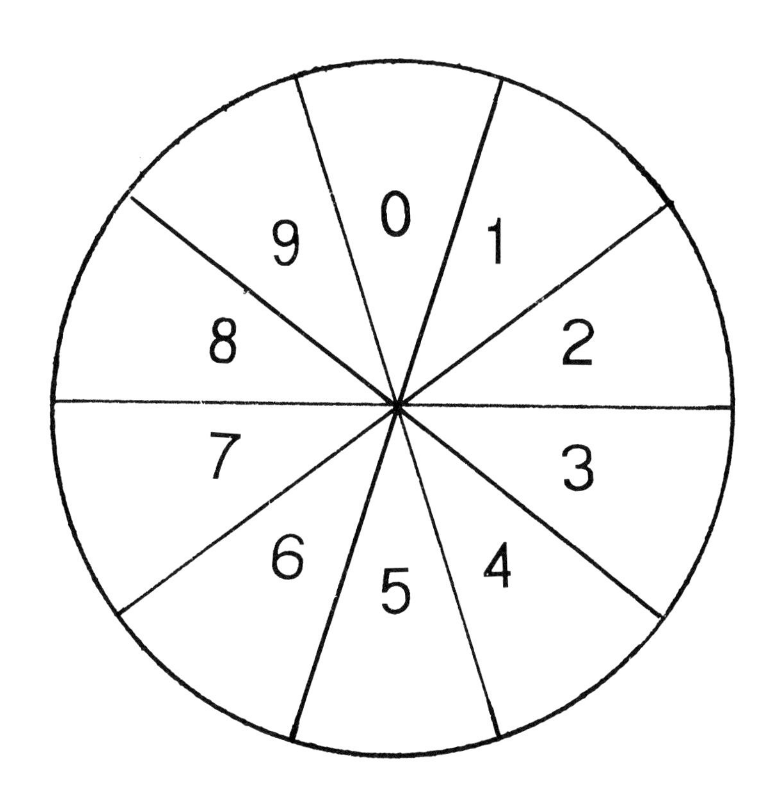

CLAIRE'S GAME

A game for 2 - 6 players
Invented by Claire who is 4.

****

This game develops ideas of equivalence. For example, a 5-rod and a 2-rod are equivalent to a 7-rod. This involves number bonds and estimation of length. It is possible for children to play this game without fully developed number concepts, as exchanges are made through comparison of length.

****

Use Cuisenaire® sheet 7 to make a staircase board for each player...

A die is needed.

In every game the winner is the first player to fill each step with one rod.

Game 1

- Players take it in turns to throw the die, and collect that number of white rods.
- The white rods are placed on the staircase.
- When one step is complete, the white rods are exchanged for a single rod which fits the space.

cont...

Game 2

- Players take it in turns to throw the die and take one rod equivalent to the number thrown. For example,

- To get a complete set, players must exchange rods during their turn.

Game 3

- Fill the staircase with rods.
- Players take it in turns to throw the die.
- They PUT BACK the number shown on the die.

Either 1.They must wait for a number to appear before removing a rod,
or 2.They can exchange rods eg. if three is thrown, they can give a five and get a two in exchange.

Game 4

- Let them invent their own game using a different shape or number pattern.

Cuisenaire® Sheet 7

JONATHAN'S TUGBOAT GAME

A game for 2 - 6 players
Invented byJonathan who is 6

A die or spinner is needed.

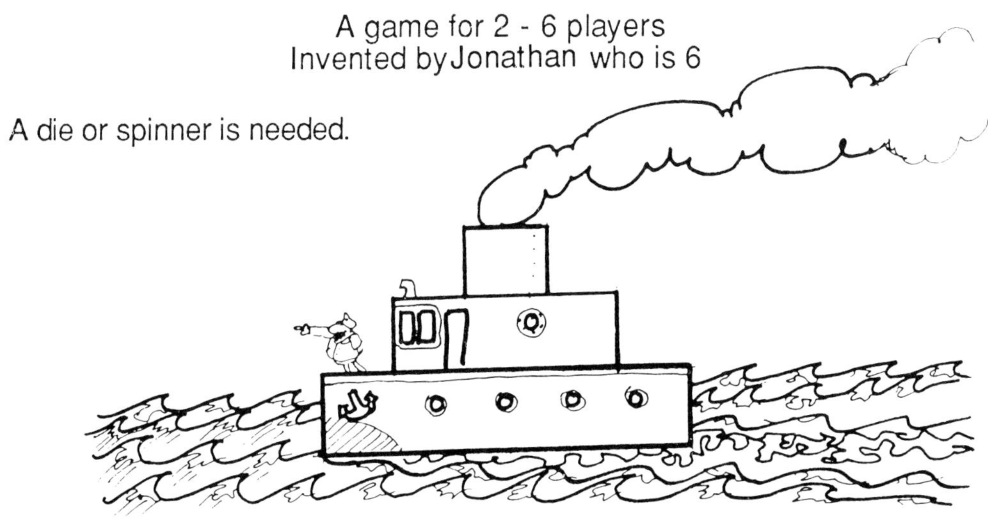

Introduction to the game.

- Ask the children to make tugboats from three rods of different lengths,
  eg.

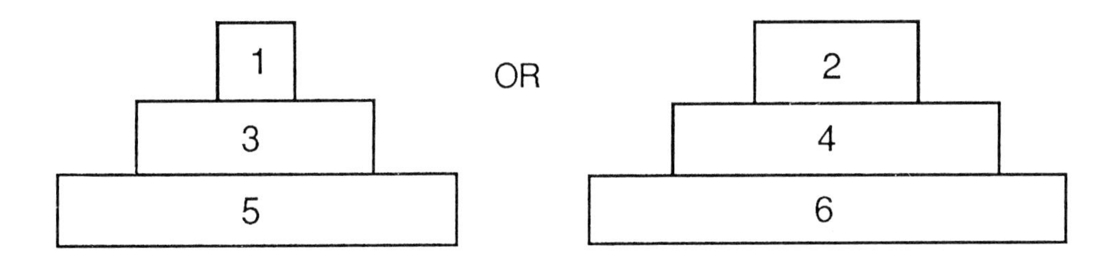

To play

- Players take it in turns to spin or throw a number and collect rods to that value.
  For example, if 5 is thrown players may take a 5-rod, a 3-rod and a 2-rod, etc.

- The winner is the first player to complete five tugboats.

It is essential that the children discover for themselves the most economical tugboat..

Variations.

1. Change the number of tugboats.

2. Specify the size of the tugboat.
   eg. 2, 5, 7.

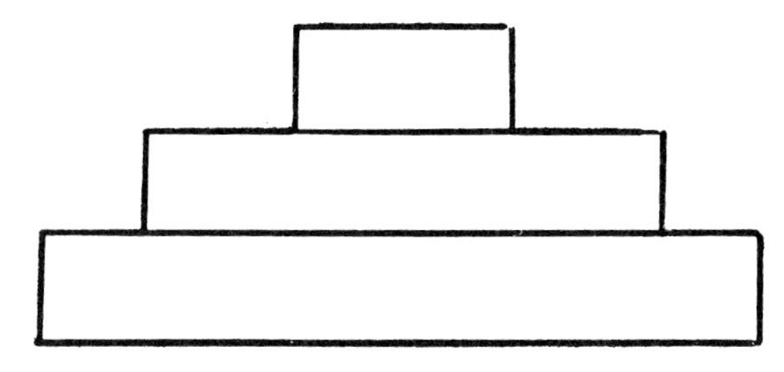

3. Specify a relationship between the rods, eg. a difference of two,

   eg. 2, 4, 6.

4. Use larger numbers and two dice.

5. Use subtraction, multiplication and division, or a combination of operations.

6. Play the game in reverse. Begin with five tugboats and give back whatever is thrown on the die.

7. Throw two dice. Use the total thrown and make a tugboat. If you cannot make a tugboat - miss a turn.

8. It is even better if children invent variations of these games or invent new games of their own.

STAIRCASES

Use Cuisenaire® Sheet 7 or 8

****

This activity generates patterns in colour and number

****

Use the staircase board with the following problem:

> Choose two colours of rod.
> You may use as many of these as you need.
>
> Can the staircase be filled using only these colours?

Further questions and investigations

1. Which two colours fill the most steps?

2. Which two colours fill the fewest steps?

3. Which two colours fill only ten steps?

4. Are consecutive pairs the best, eg 3 and 4?

5. Use only one colour eg. green.
   Which stairs can be filled?

6. Are there three rods which can be used to fill the whole staircase?

7. Can you predict which rods will work?

Cuisenaire® Sheet 8

CARRIAGES

A game for 2 players.
Cuisenaire® sheet 9

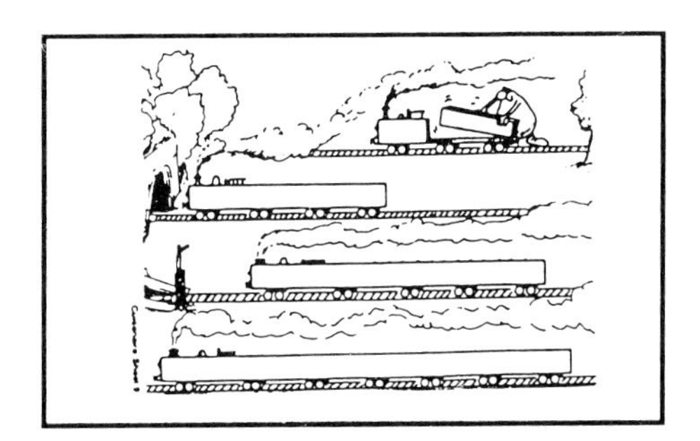

To play

Use white, red or green rods as carriages, and any one of the trains on Cuisenaire® sheet 9.

- Players take it in turns to put a carriage on the train, starting at the front.

- The player who adds the last carriage wins the game.

- Discuss the possibility of a strategy that ensures a win?

Variations

1. The player who adds the last carriage loses.

2. Use a different set of rods for carriages.

Extensions

Cuisenaire® sheet 9 can also be used in estimation and counting activities.

For example:

1. What is each train worth?

2. Is there a colour of carriage that will fill all the trains? If not try two colours.

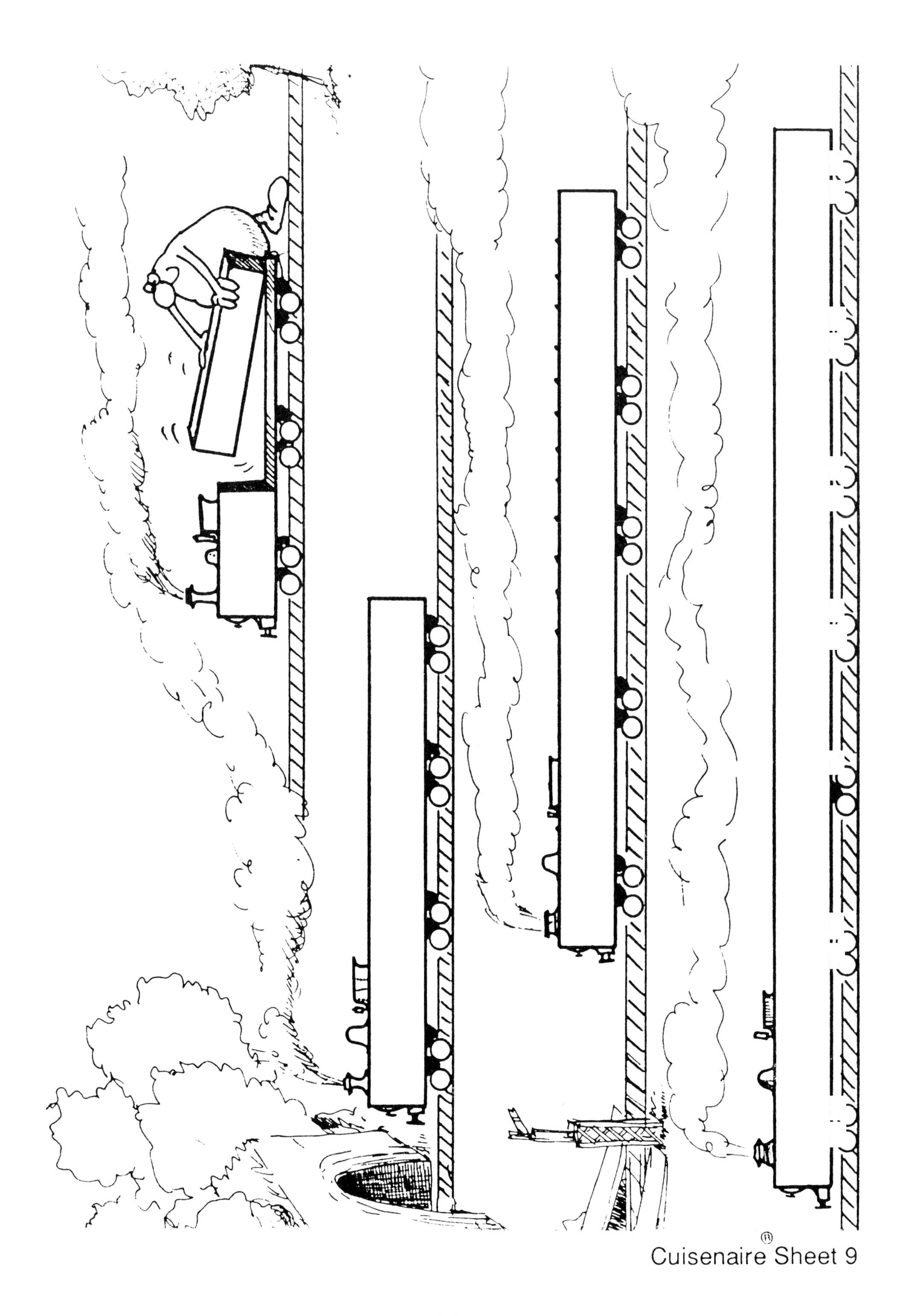

Cuisenaire® Sheet 9

WINDOWS

Cuisenaire® Sheet 10

****

The aim of this activity is to give children a variety of spatial experiences including the conservation of area.

****

This activity can develop through the following questions and challenges:-

- How many different curtains can be made by covering the windows with rods?
- Which is the longest rod that can be used?
- Which is the shortest rod that can be used?
- Make trains from the rods in each curtain.
  What do you notice?
- Can 9 windows be filled using 4 rods?
  Is there more than one way of doing it?
- Can they be filled using 7 rods?
- Can a curtain be made using only 2 colours?
- What is the fewest number of rods that can be used?
- What is each window worth?
- What is the block of flats worth?
- How many windows?
- How many windows for 100?
- Which numbers will make windows?

...... and others - that children may ask themselves.

<u>Extensions</u>

1. Use Cuisenaire® sheets 11 or 12, ask similar questions and compare the answers.
2. Make 2 x 2 windows: How many rods to cover 1?
   How many rods to cover 2?
   How many rods to cover 3?
   Colour the results in on the hundred square.(Find one at the back of the book)
   What do you notice?
3. Use different windows. (Not necessarily squares!)

Give each window some curtains

Cuisenaire® Sheet 10

Cuisenaire® Sheet 11

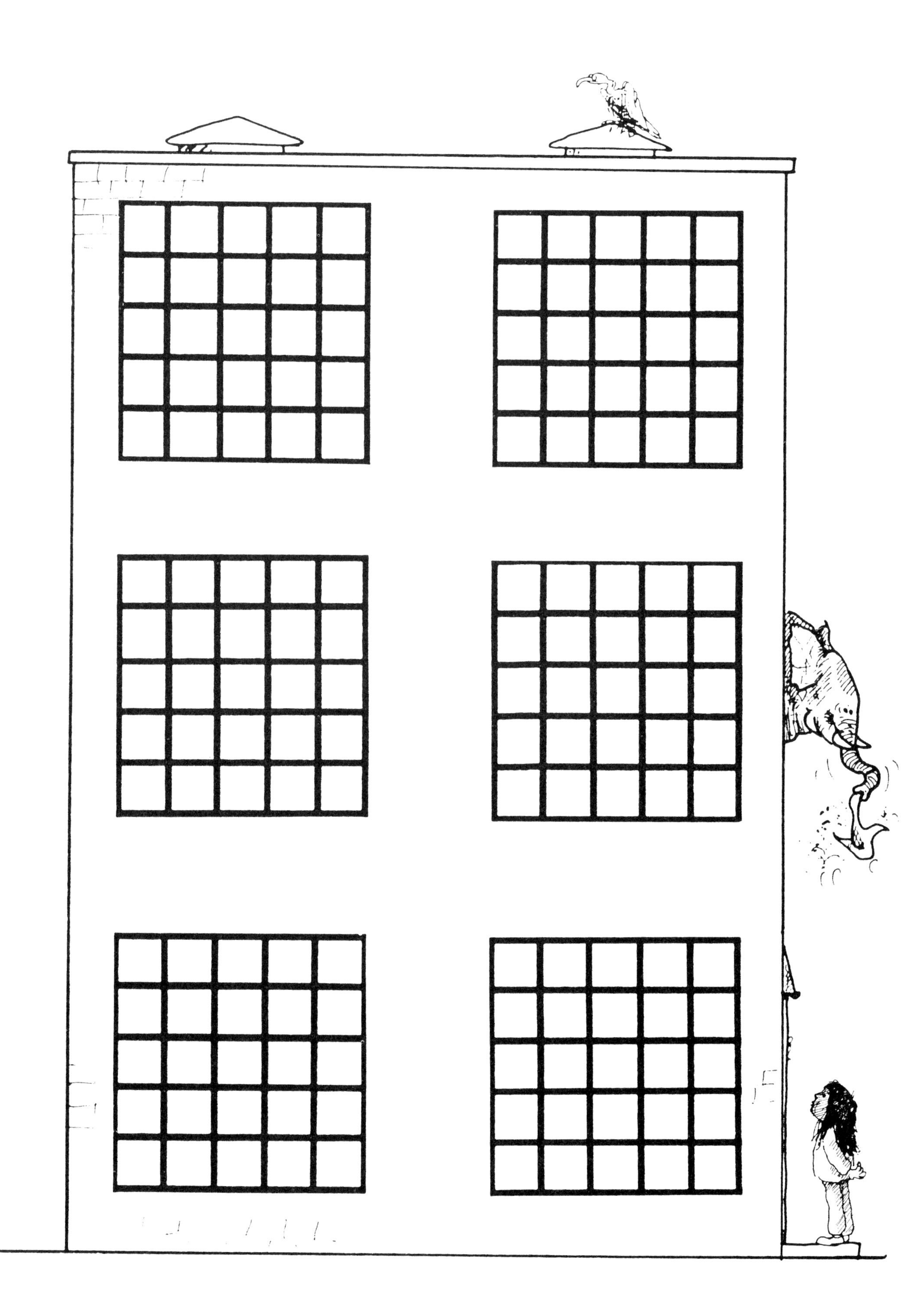

THE WORST GAME I KNOW

Cuisenaire® sheet 13

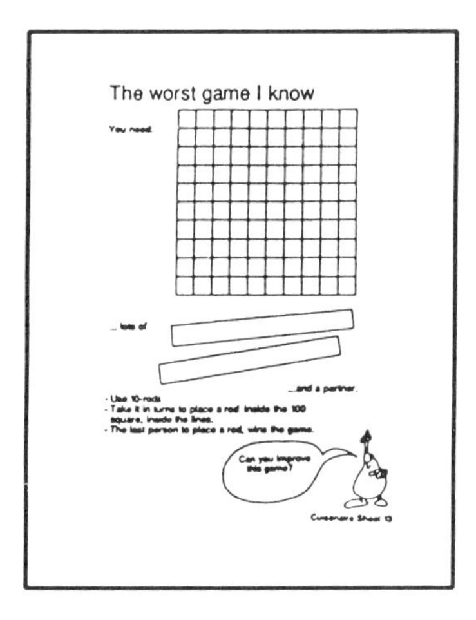
The worst game I know

You need

... lots of

...and a partner.

- Use 10-rods
- Take it in turns to place a rod inside the 100 square, inside the lines.
- The last person to place a rod, wins the game.

Can you improve this game?

Cuisenaire Sheet 13

****

The value of this activity lies in the lack of direction children are given, and in the freedom they have to create their own games.

****

Children are allowed/encouraged to invent their own games. Cuisenaire sheet 13 may be used as a game board on which children invent their games by modifying the awful example suggested. The first games may be similar and possibly mundane, but as they become used to the idea, children will become more inventive.

The worst game I know

You need:

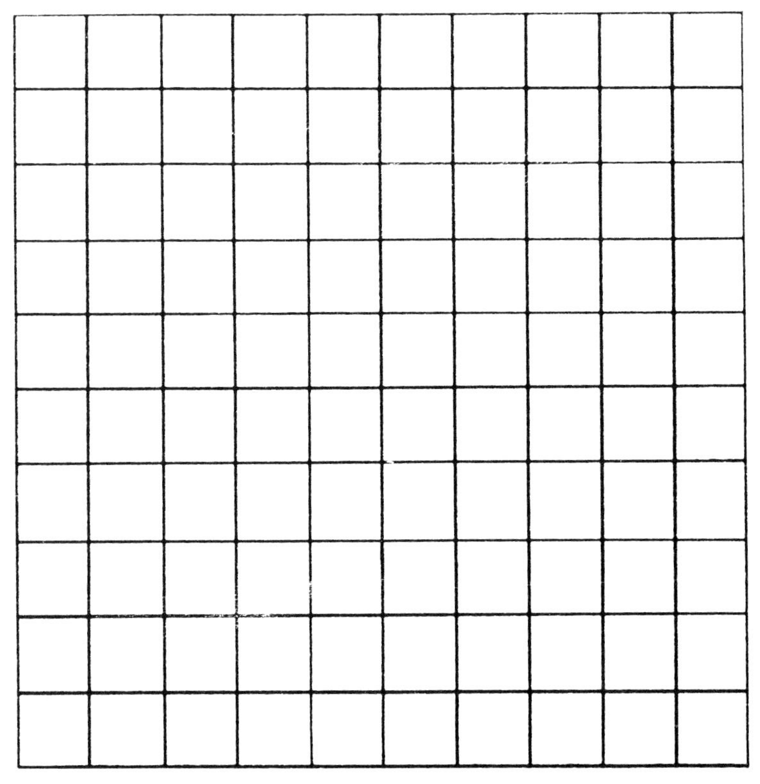

... lots of

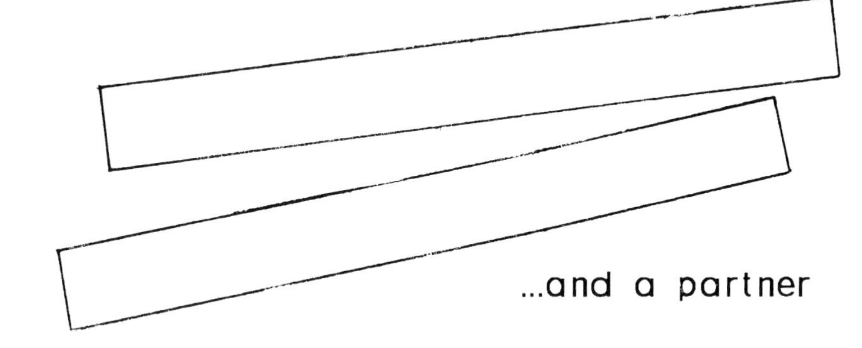

...and a partner

- Use 10-rods
- Take it in turns to place a rod inside the 100 square, inside the lines.
- The last person to place a rod, wins the game.

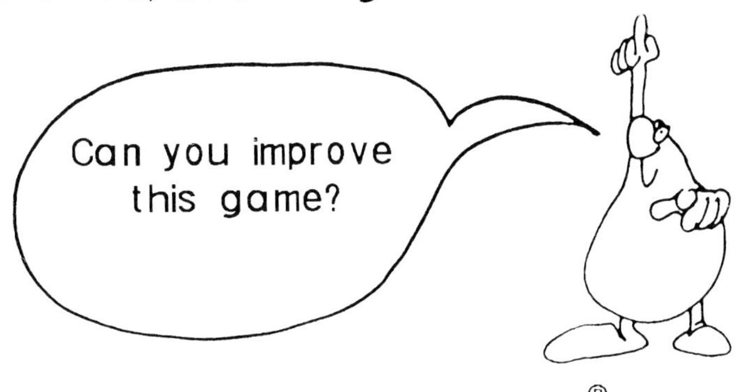

Cuisenaire® Sheet 13

GUESS MY RULE

Games for two or more players

****

These activities develop mental arithmetic and the ability to recognise patterns in number

****

Both games can be played:-

- as a game for two players
- as a game for two teams
- as a game where one child (or maybe the teacher) challenges the rest of the group.

<u>Game 1</u>

One player thinks of a rule, for example 'The rod which is the next size'

The other player, team or group, must guess that rule.

<u>To play:-</u>

- The challenger(s), choose a rod.

eg.

- The Rule-maker takes out the rod which fits the rule, ie. is the next size, and places it next to the first rod.

eg.

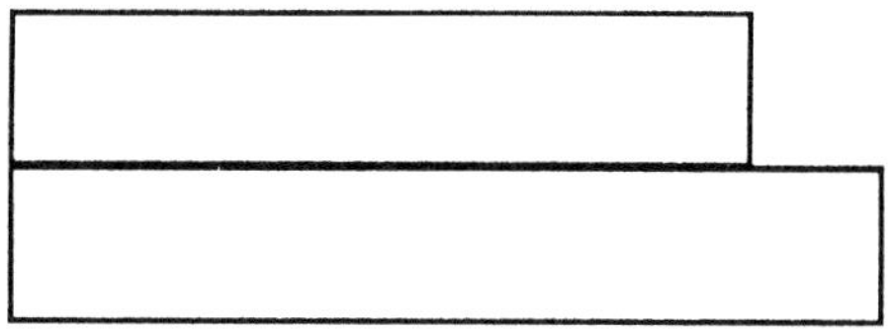

- This continues until the challenger(s) want to 'Guess the Rule'. Whoever guesses correctly makes the next rule.

cont...

Other rules might include

- The rod (or set of rods) which make a train twice as long as the first rod.
- The rod which is two less than the first.
- The first rod plus the second rod make ten.
- Etc.

Game 2

One child is chosen as the 'Magician' and sits behind a screen. Alternatively, sits with his/her hands and some rods hidden behind a screen.

As in game 1, the magician must choose a rule.
eg. 'The rod which is one rod smaller'.

- The challenger(s) choose a rod and pass it behind a screen to the magician.
- The magician exchanges the rod for the rod which is worth one less, than the one she was given. She hands the exchanged rod back to the challenger.

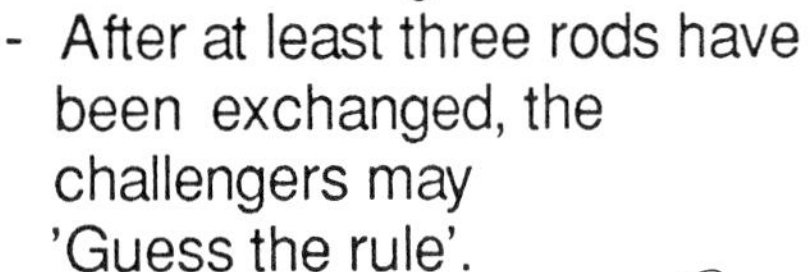

- After at least three rods have been exchanged, the challengers may 'Guess the rule'.
- Whoever guesses correctly becomes the magician for the next game.

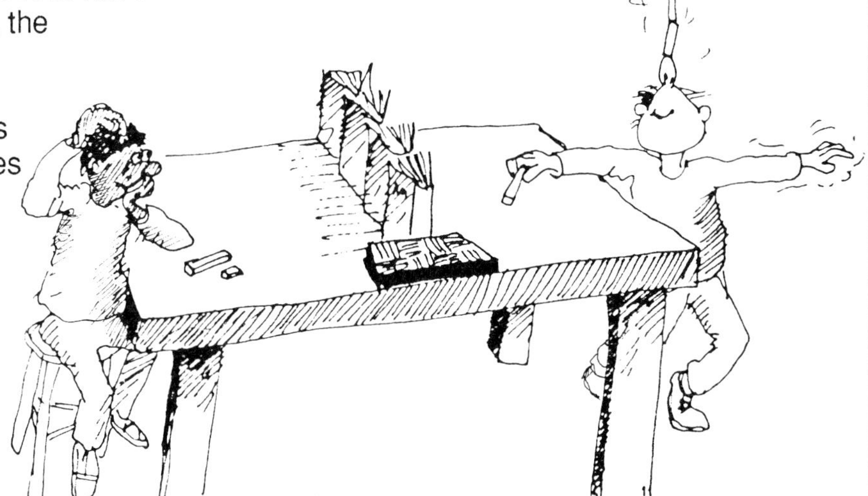

THE GATE GAME

A game for a group of children

To play

- One child is chosen as the Gatekeeper.
- The gatekeeper invents a rule that will open the gate, eg. 'Is more than 5'.

- The rest of the group take it in turns to choose a rod each, that they hope will get them past the gatekeeper.

- They ask if they will be allowed to go through the gate.

- Those children who are holding a rod longer than the 5-rod, will go through the gate.

- Once everyone has either been allowed through the gate or not, children are asked to work out what the gatekeeper's rule was. The child who gives the correct solution becomes the gatekeeper for the next game.

This activity is a good introduction to tree diagrams. Cuisenaire® sheets 14 - 17 are given for this purpose.

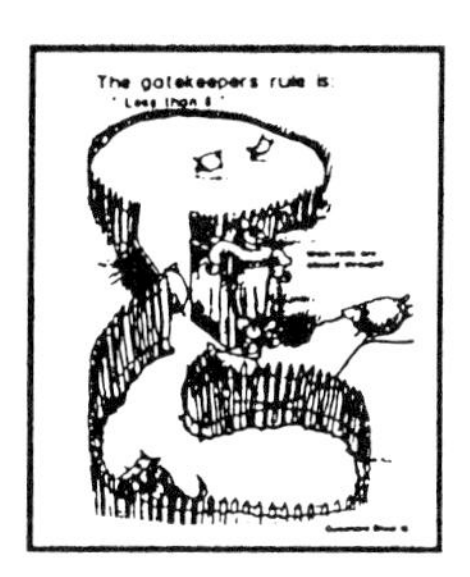

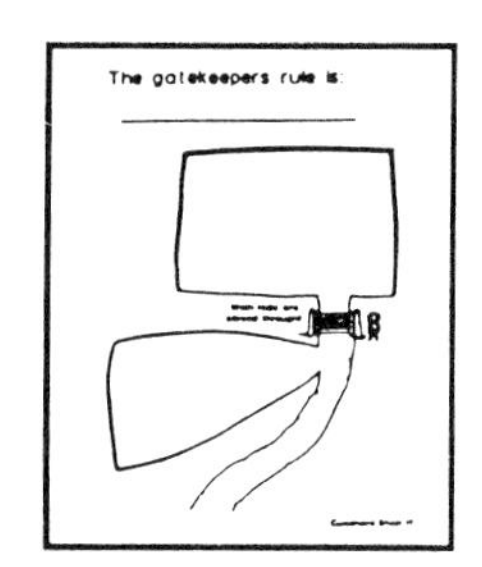

The gatekeepers rule is:

"Longer than the pink rod."

Cuisenaire® Sheet 14

The gatekeepers rule is:

" Less than 8 "

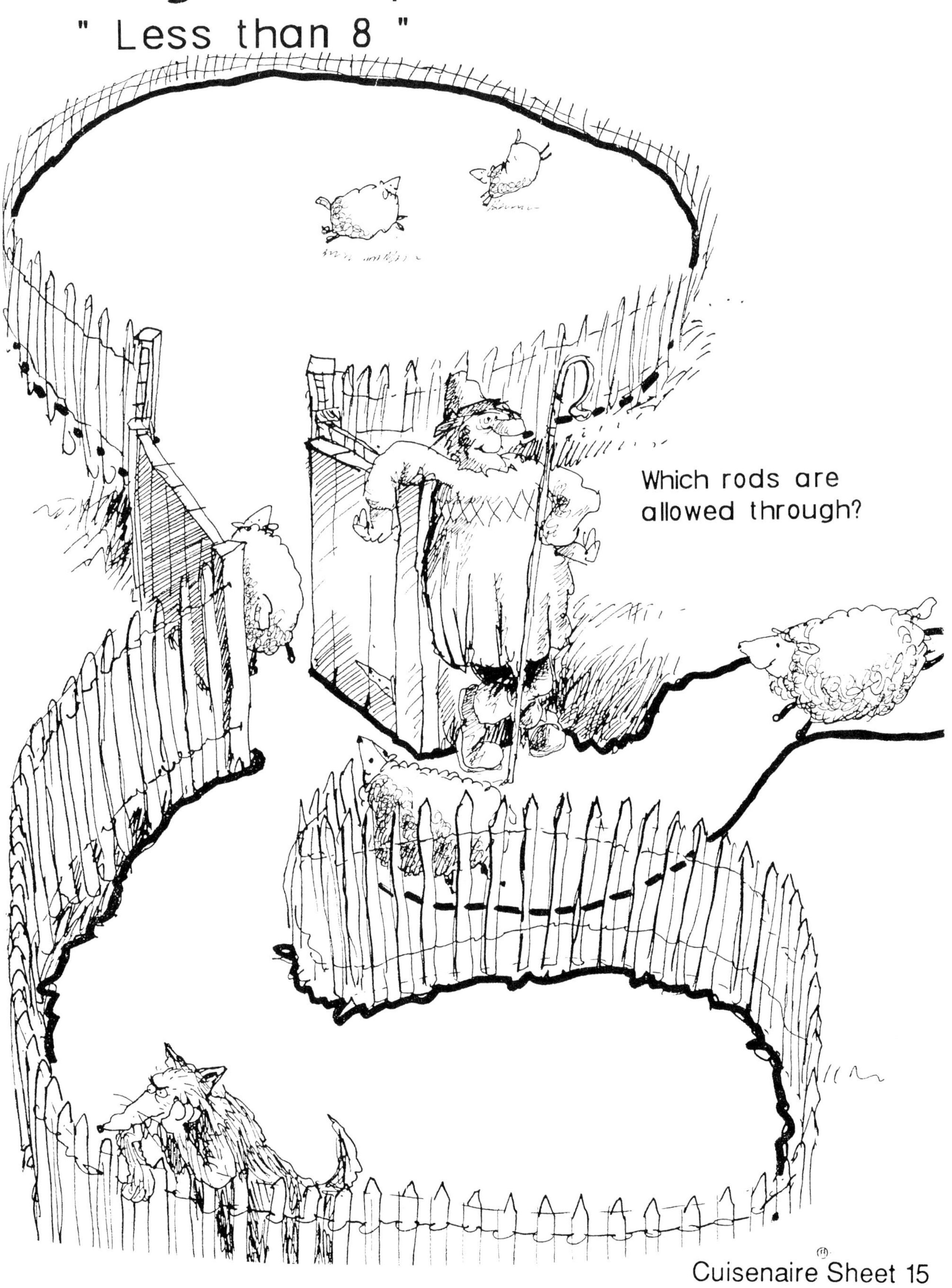

Cuisenaire® Sheet 15

The gatekeepers rule is:

" Red rods will fit beside it exactly "

Cuisenaire® Sheet 16

The gatekeepers rule is:

____

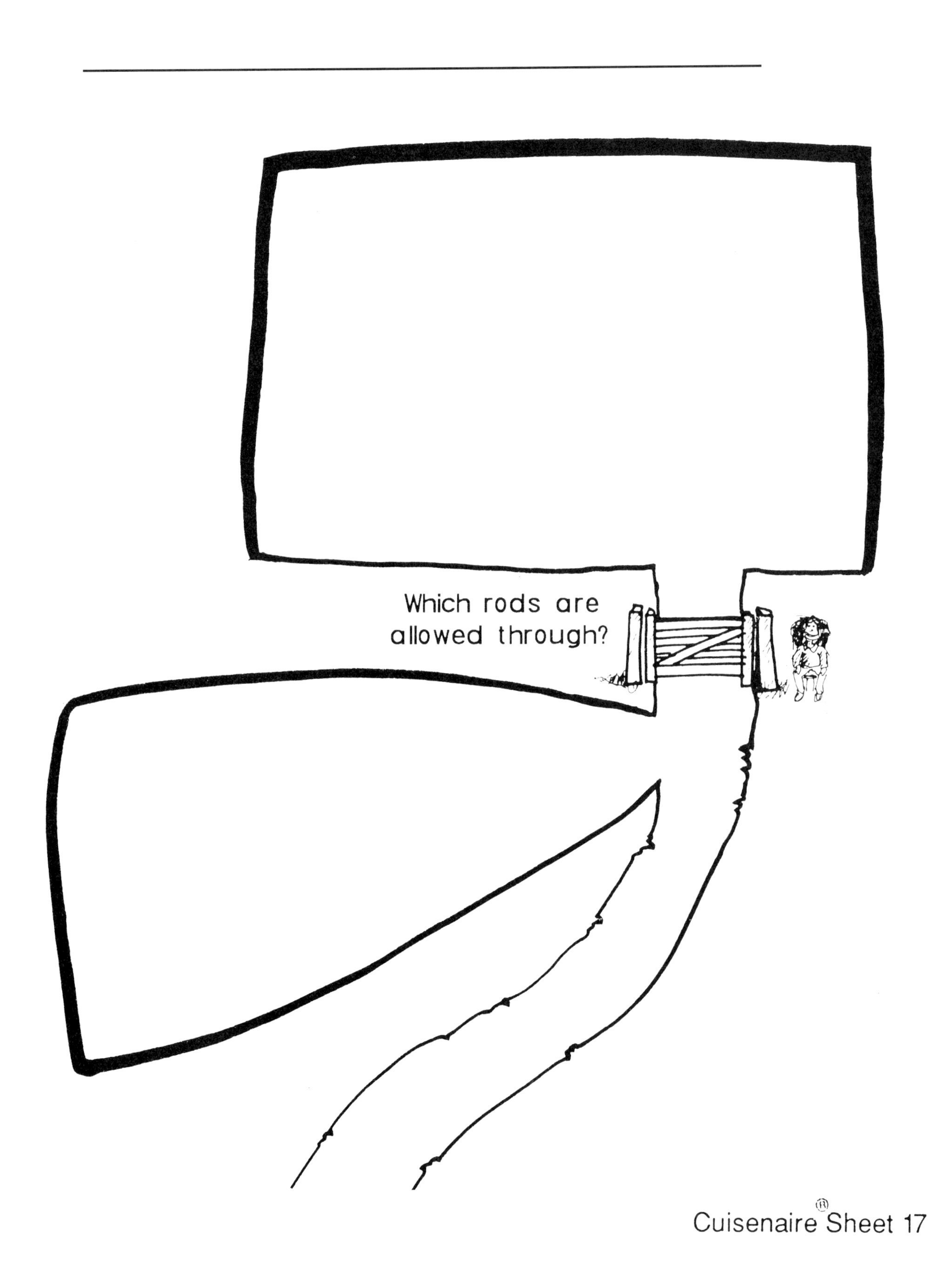

Cuisenaire® Sheet 17

COVERING SPACES

Cuisenaire® sheets 18-21

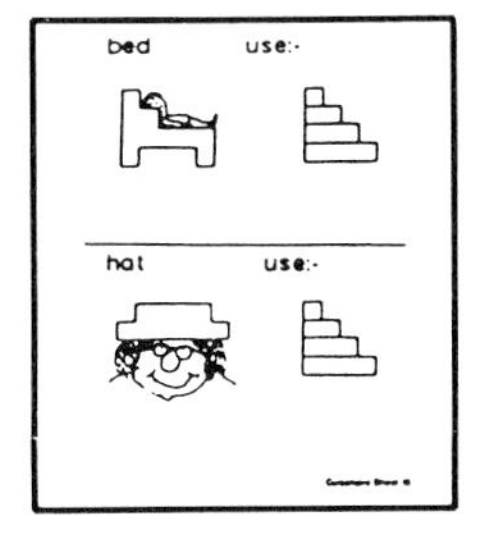

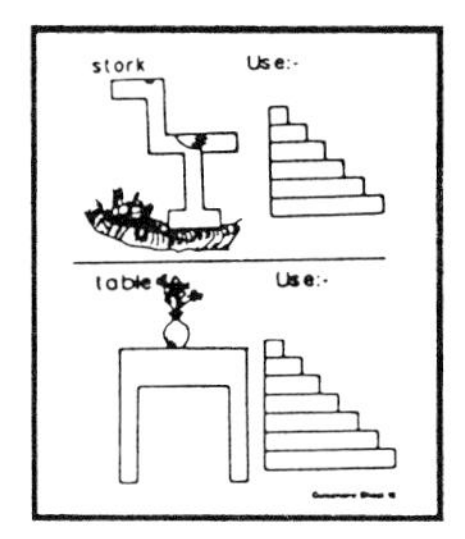

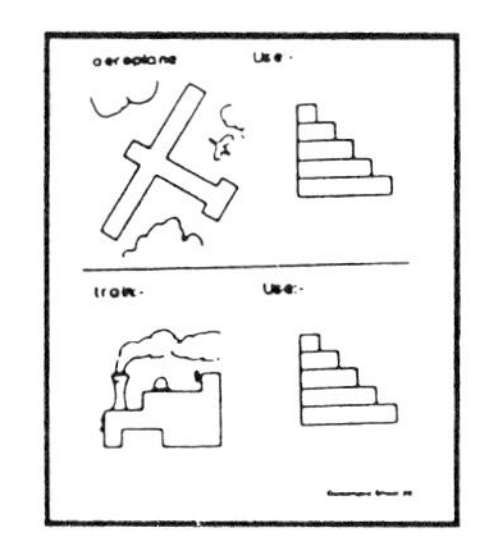

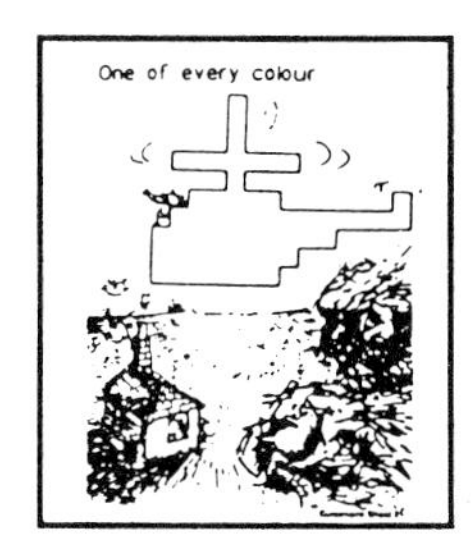

****

These activities develop spatial and/or numerical concepts.

****

The pictures on sheets 18 - 20 can be cut into individual worksheets, or copied onto card and laminated to be used as workcards.

These pictures can be filled with the rods shown. Before asking children to do this, we suggest the following activities:-

1. Let the children guess how many white rods will fill the space.
2. Put 4 or 5 anywhere in the shape and let them revise their guess.
3. Fill the shape. How good was their guess?
4. Now fill the picture using any rods.
5. Use specified rods to fill the shape.

<u>Extensions:</u>

1. Use two colours to fill it.
2. A game for two players:-
   Take it in turns to put a rod onto the picture.
   The player who completes the picture by placing the last rod LOSES the game.
3. Use as many rods as possible.
4. Use as few rods as possible.
5. Choose one set of rods from the above activities and use them to make three new pictures.

bed use:-

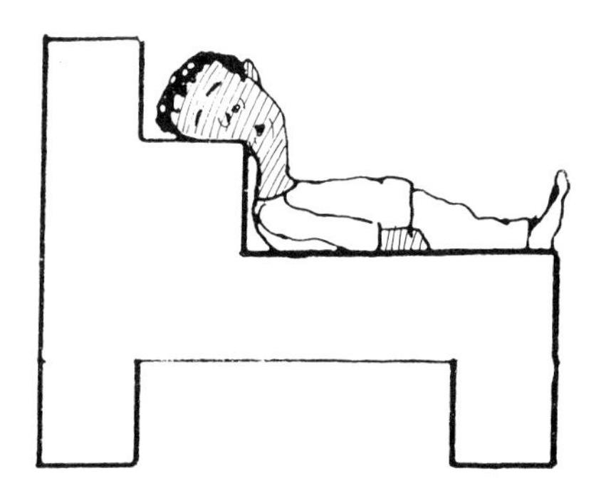

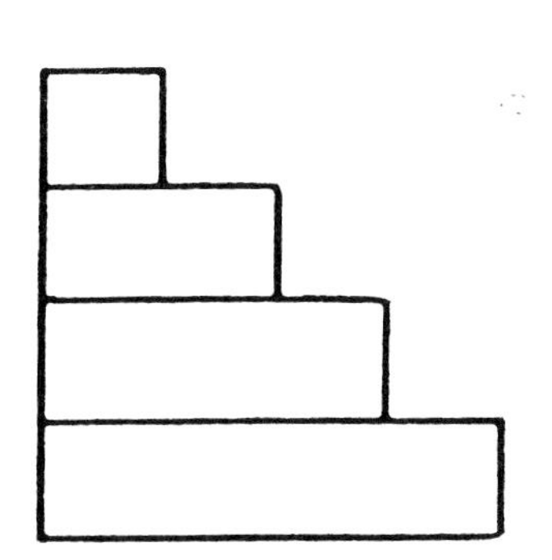

---

hat use:-

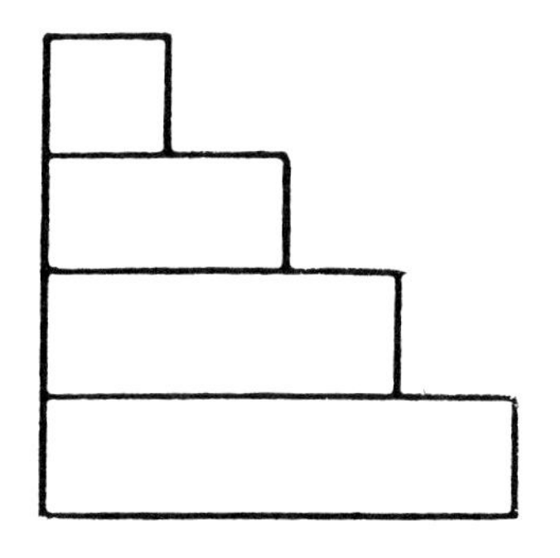

Cuisenaire Sheet 18

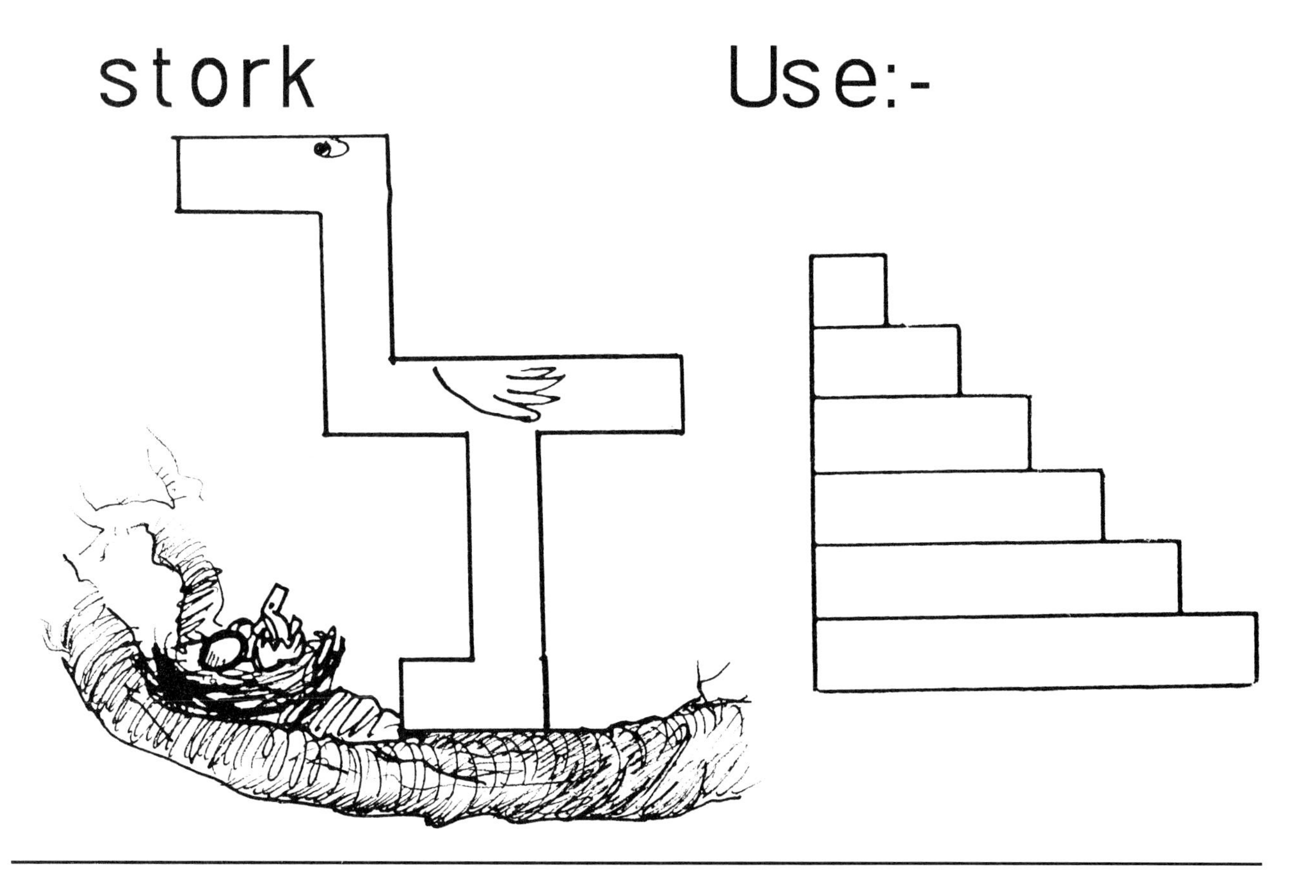

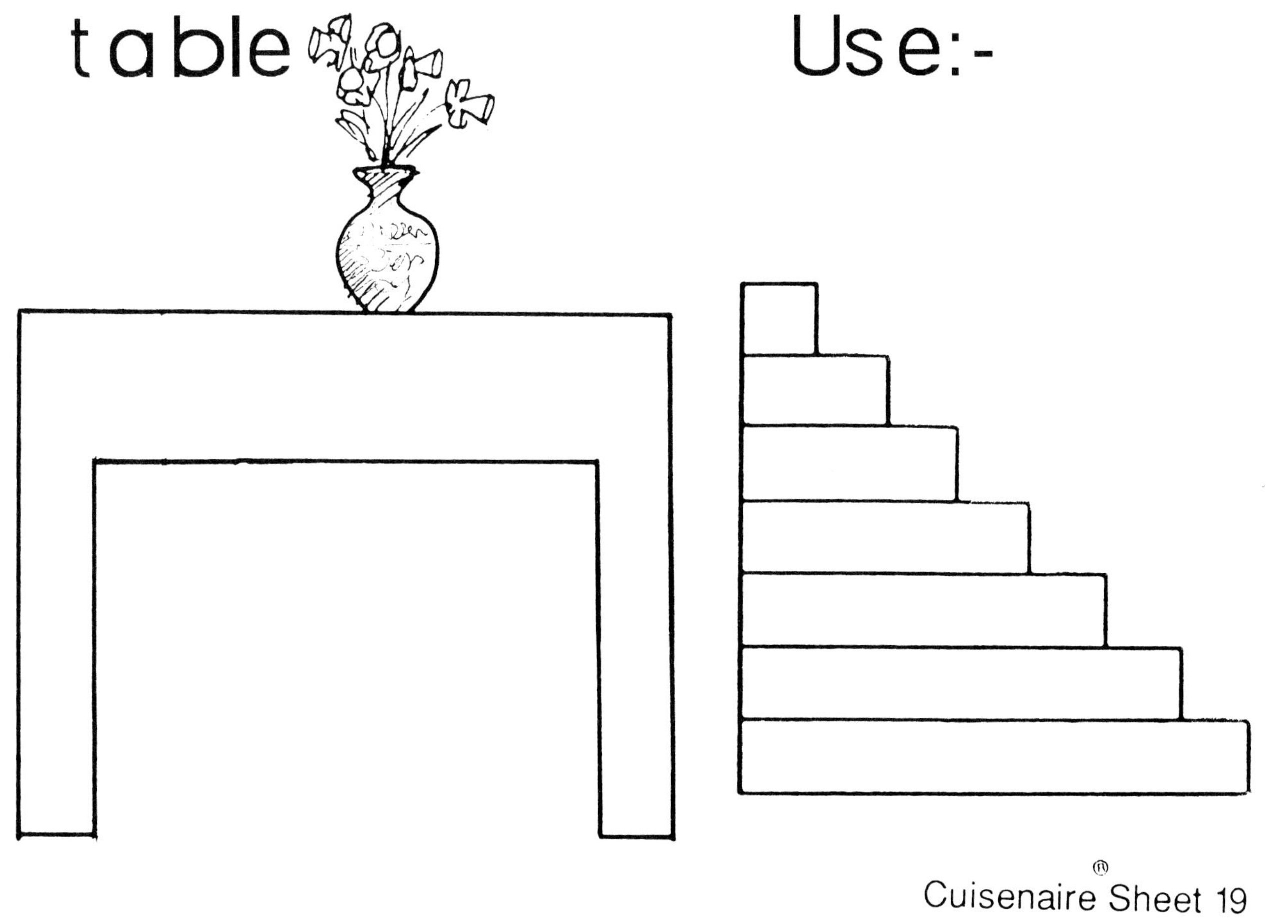

Cuisenaire® Sheet 19

aeroplane

Use:-

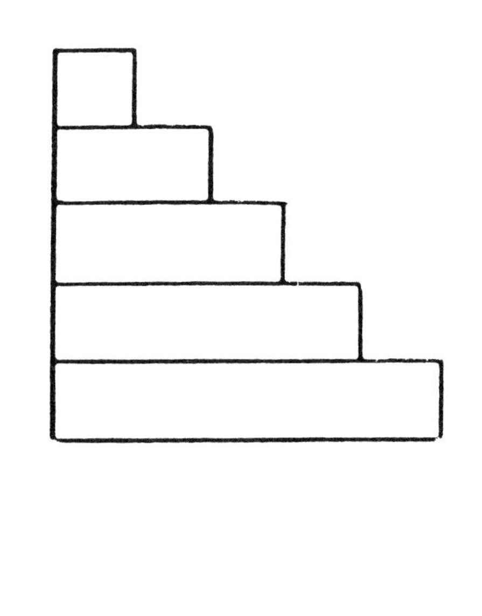

train:-

Use:-

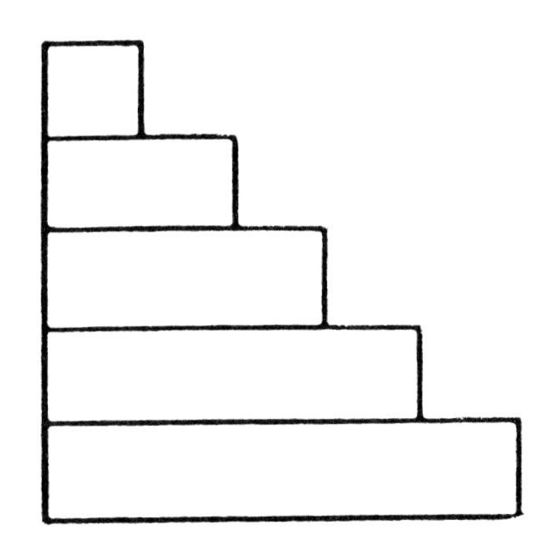

Cuisenaire® Sheet 20

One of every colour

Cuisenaire® Sheet 21

INTRODUCTORY ACTIVITIES

Rods offer a unique and infinitely flexible model of our number system, as well as a measuring tool for length, area and volume. They should not be the only equipment children use, but part of an armoury of equipment available to children in the development of mathematics.

Two basic ideas are pivotal to the use of rods. These are:

i) An association between colour, length and number.

ii) The idea of a train i.e. that two or more rods joined together are equivalent to other rod(s).

eg.

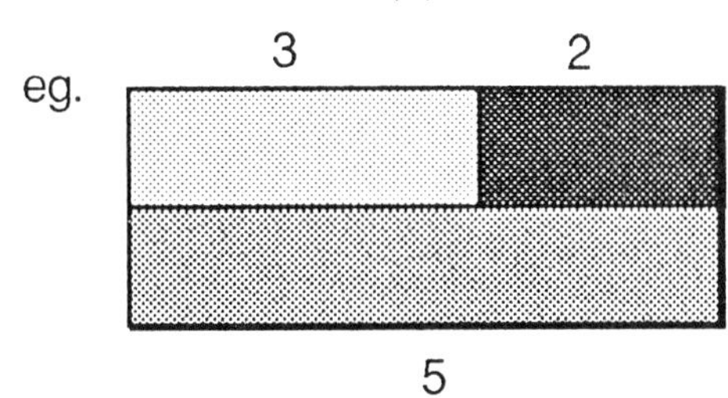

or

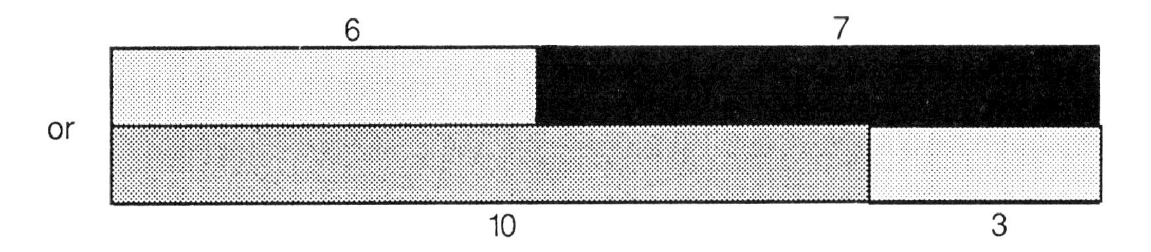

The following activities develop one or both of these ideas.

Cuisenaire® rods are available from: The Cuisenaire Company
11 Crown Street
Reading
RG1 2QT

COLOUR, LENGTH AND NUMBER

Match the rods

Cuisenaire® sheet 22

Actiivities

i) Match the rods to the spaces and record the results by colouring the worksheet.

ii) If the white rod is worth 1, find the value of each rod.

iii) Find the total value of all the rods on the worksheet.

Hands behind back

Activiites

i) 'The Three Bears' story at the beginning of this section.

ii) Children take one each of a set of rods e.g. 1,2,3,4,5 and hide them behind their backs.
Ask them to find a rod identified by colour or number without looking.
e.g. "Find the yellow rod" or "find the 3-rod"
These are two slightly different activities.

1. They put the rod in front of them.
Before the final rod is brought out, ask the children if they know what is still behind them.

2. Rods are replaced behind their backs.

'Guess It', 'Claire's Game', 'Tugboats'.

These activities can be found in the main section on Cuisenaire® Rods.

Trains

Cuisenaire® sheets 23 - 25

Ask the children to:-

i) Find a single rod to fill the shape at the top of the wall.

ii) Make as many trains as possible that are the same length as that rod.
- How many one - colour trains are there?
- How many two - colour trains?
Etc.

iii) Beside each train children can record the 'sums' they have made.

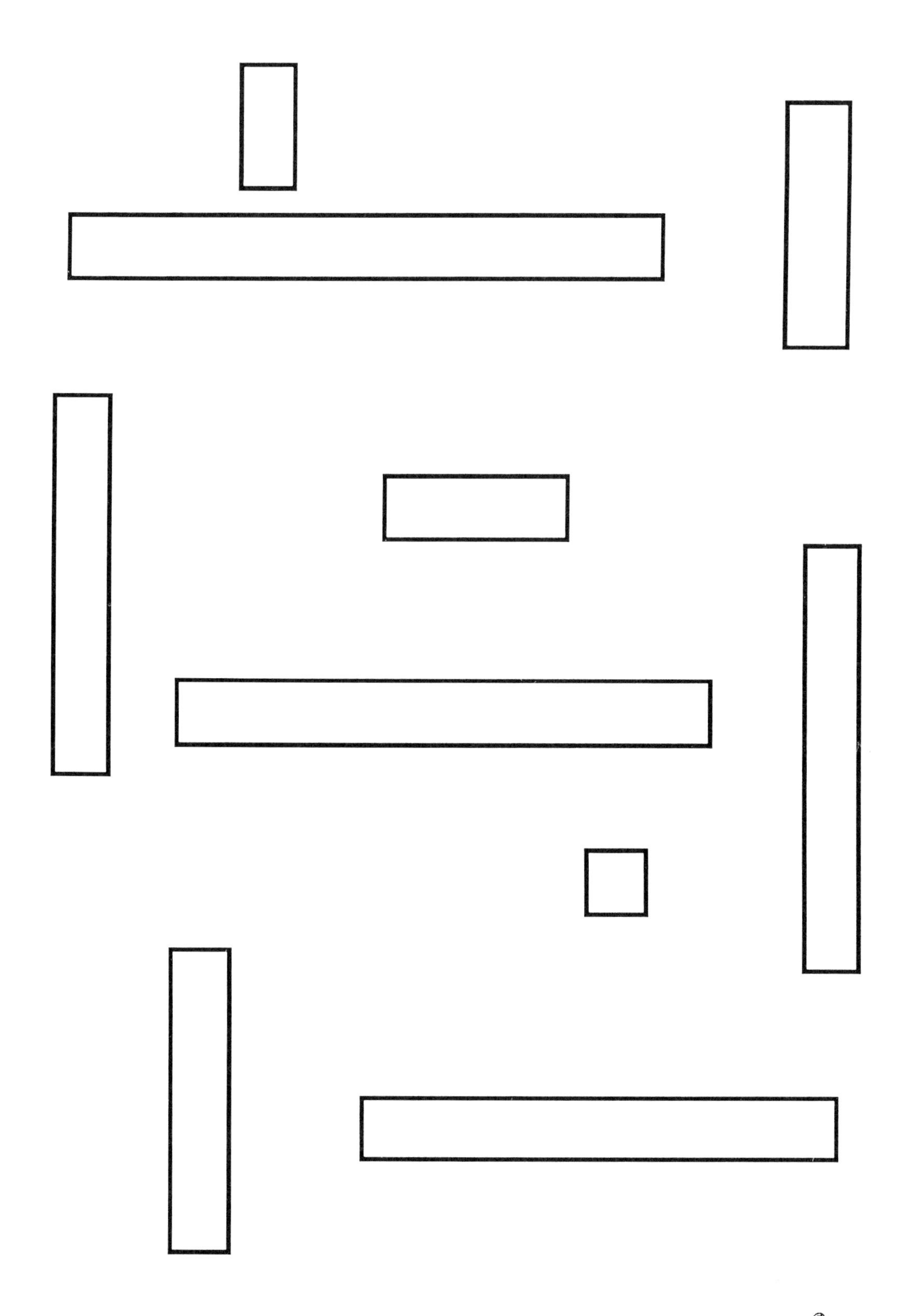

Cuisenaire® Sheet 22

Make trains

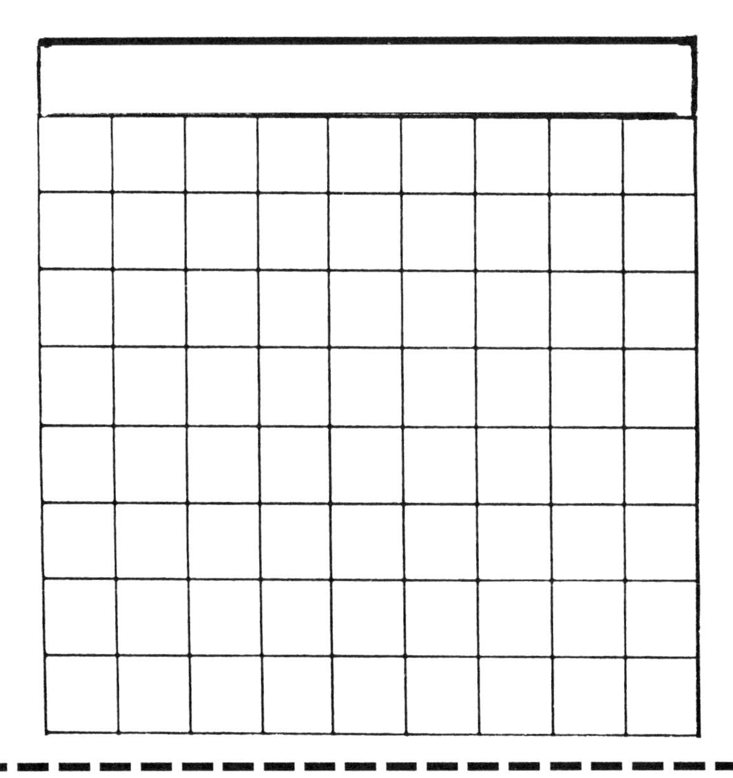

Make trains

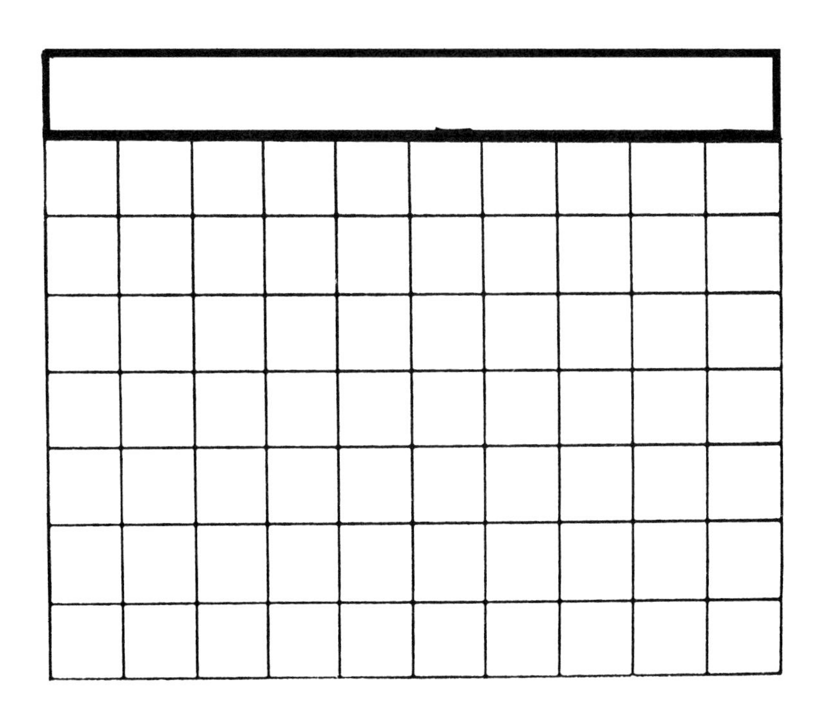

Cuisenaire® Sheet 23

Make trains

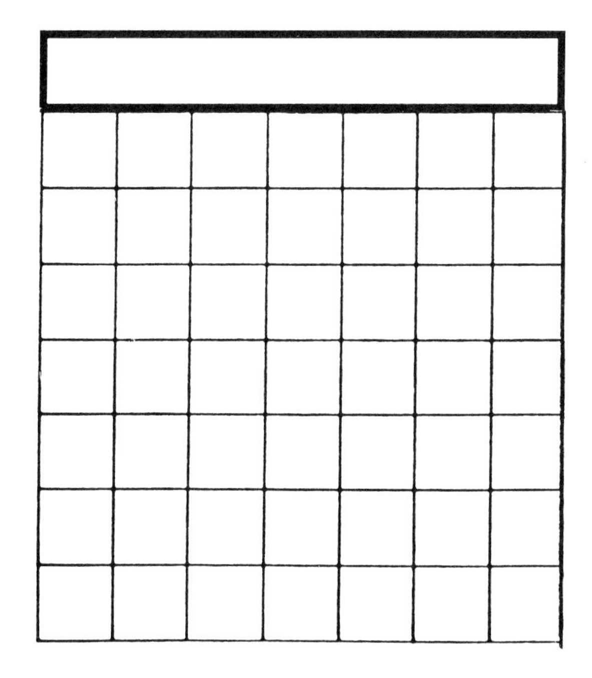

Make trains

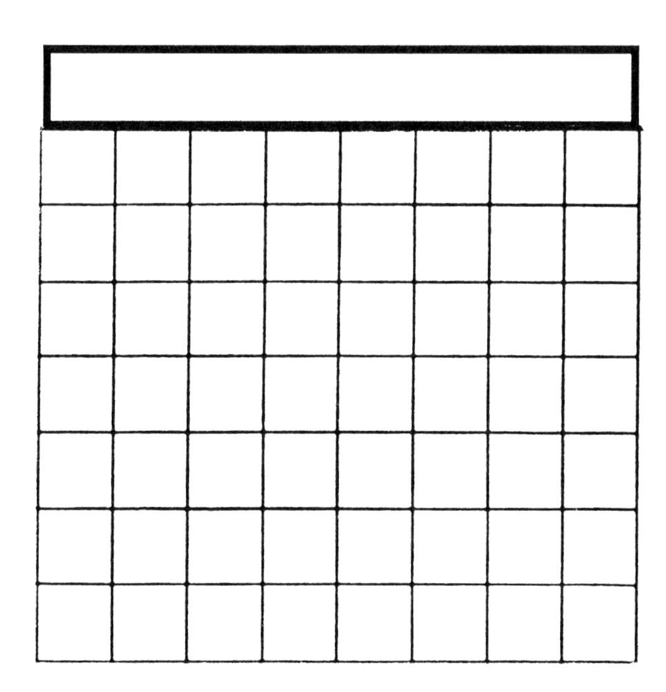

Cuisenaire Sheet 24

Make trains

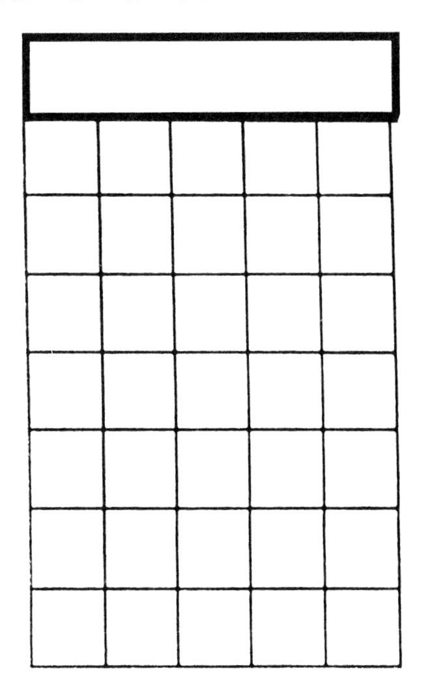

Make trains

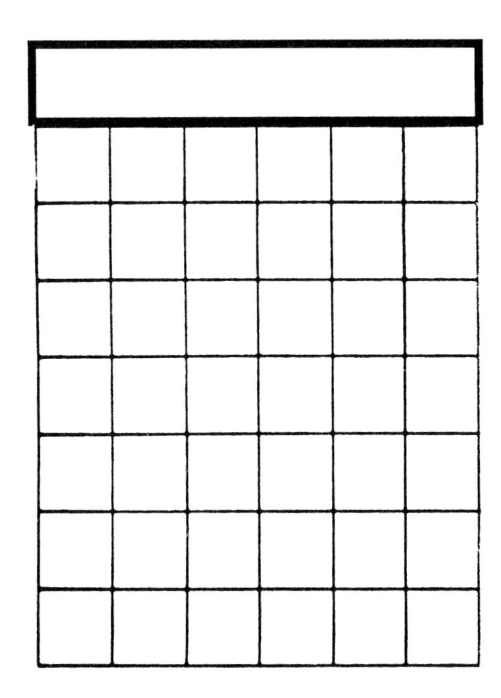

Cuisenaire® Sheet 25

COLOUR, LENGTH, NUMBER AND TRAINS

My Rod Book

Cuisenaire® sheets 26 - 29

These sheets are designed to be made into a booklet.

Children can find the value of the pictures either by addition or making trains.

Pictures

Cuisenaire® sheets 30 and 31

These sheets can be made into small individual worksheets, workcards or booklets.

Activities

Find a rod for each space on the picture.

- How many rods were used?
- Find the value of the picture by addition or making a train.
- Record by colouring.
- Use the same rods to make new pictures.

Patterns

Cuisenaire® sheets 32 and 33

These sheets can be made into small individual worksheets, workcards or booklets.

- match the rods and place them on top of the patterns.
- continue the pattern.
- copy the pattern.
- repeat the pattern using different colours.

If they are familiar with the numerical values of the rods students can be asked to find the numerical patterns associated with the pictures.

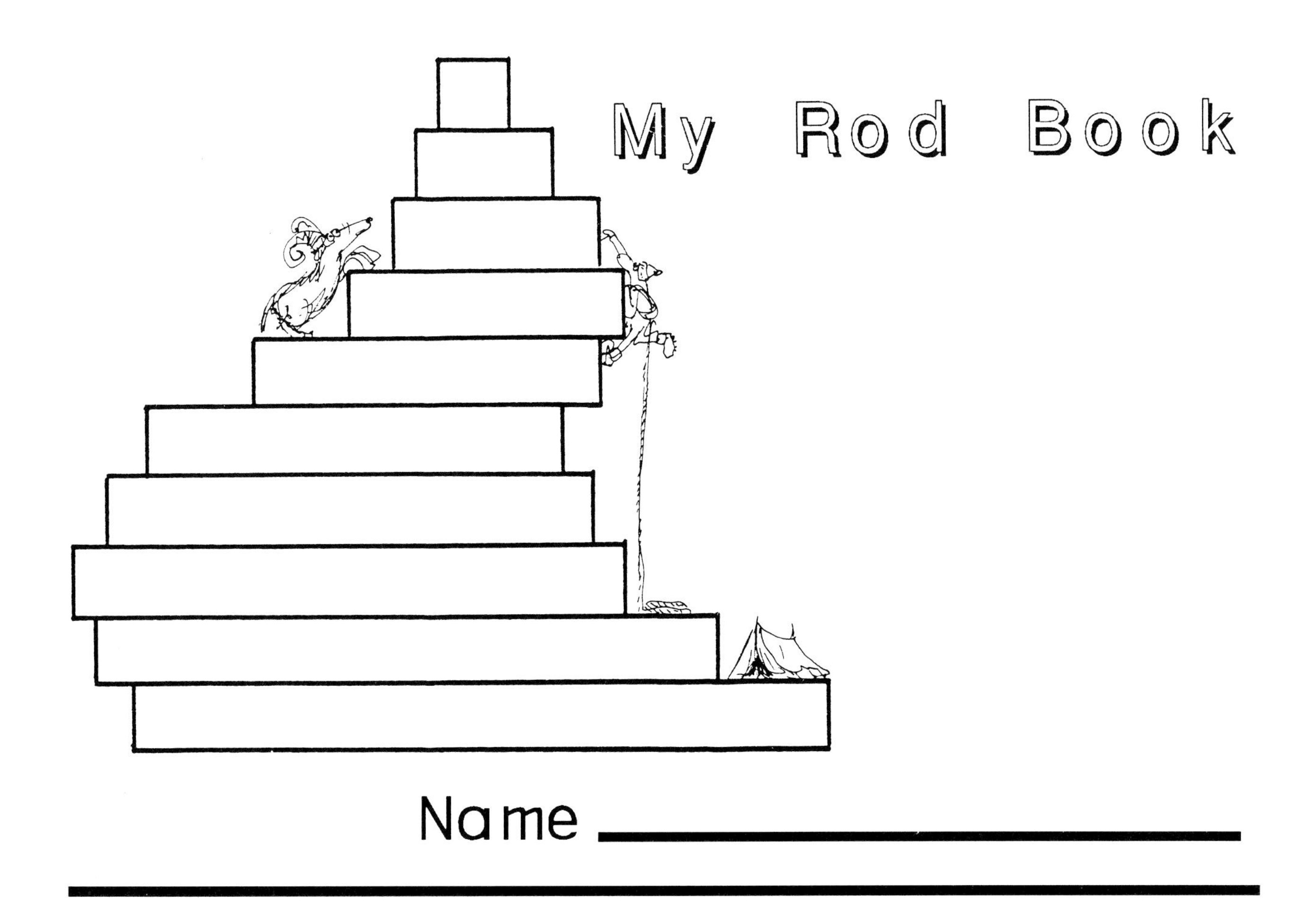

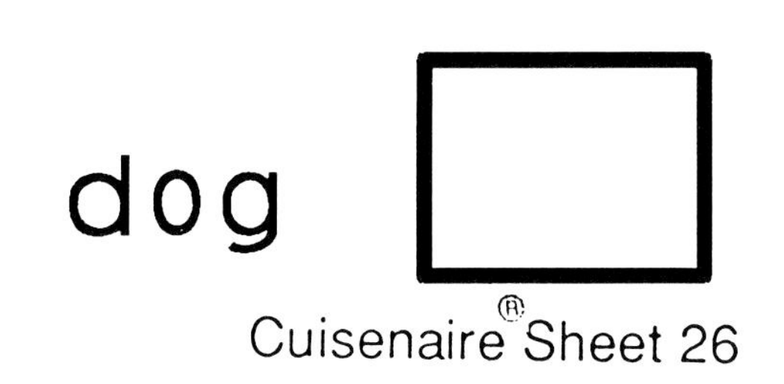

Cuisenaire® Sheet 26

lady

table

Cuisenaire® Sheet 27

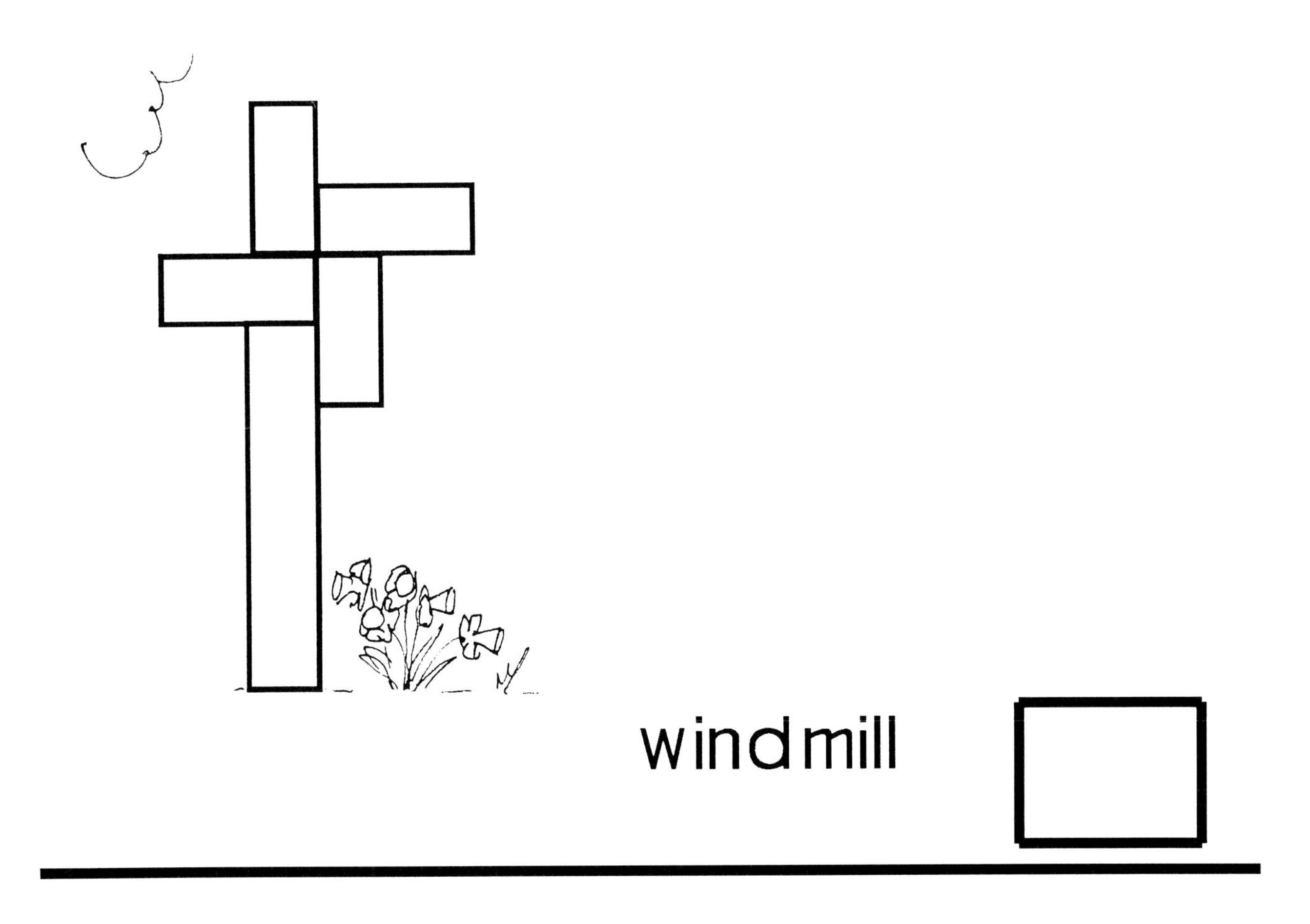

windmill

flag

Cuisenaire® Sheet 28

chair 

Make your own picture.
Draw it.

Cuisenaire® Sheet 29

Swan

Camel

Cuisenaire® Sheet 30

Face

Tree

Cuisenaire® Sheet 31

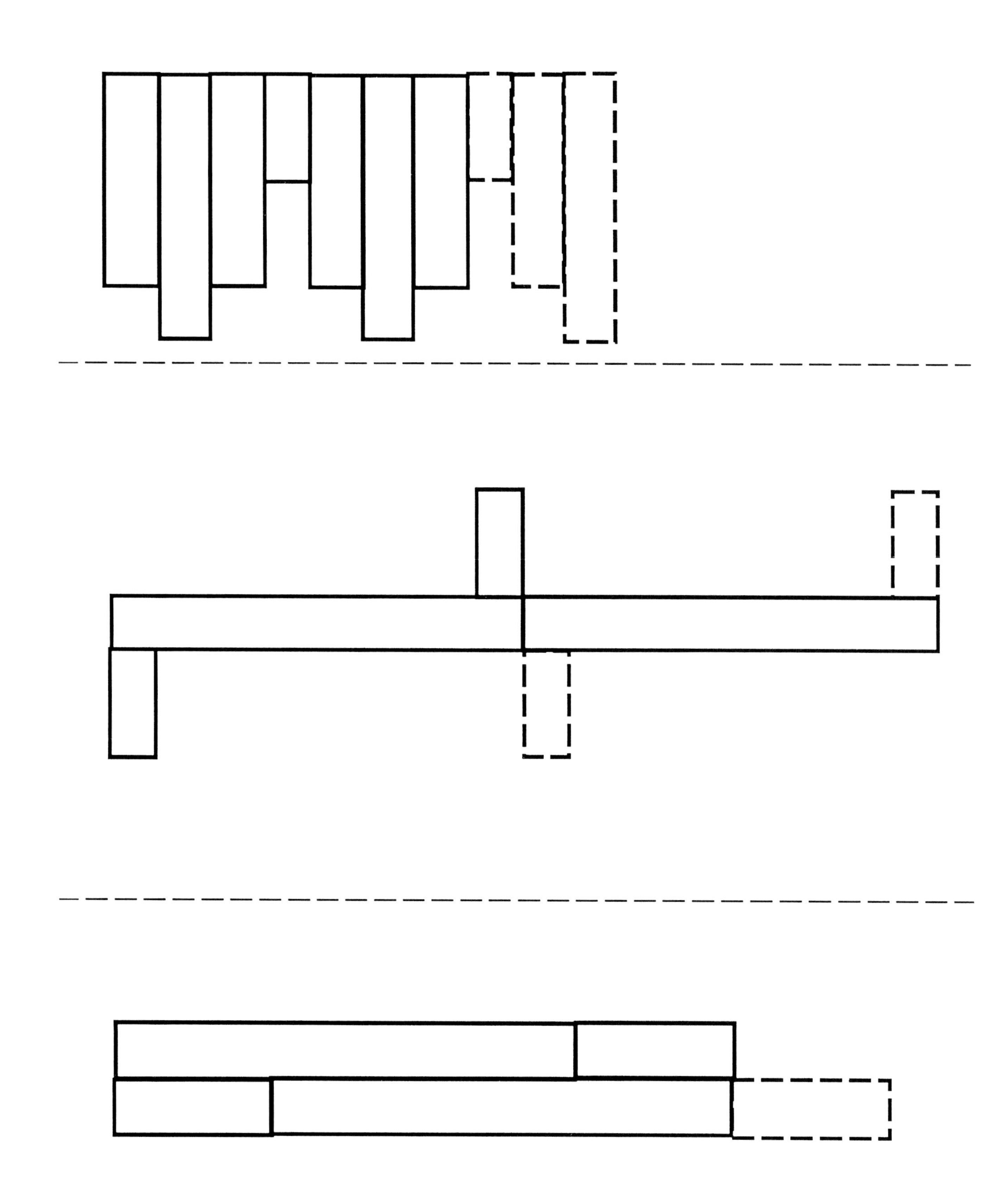

Cuisenaire® Sheet 32

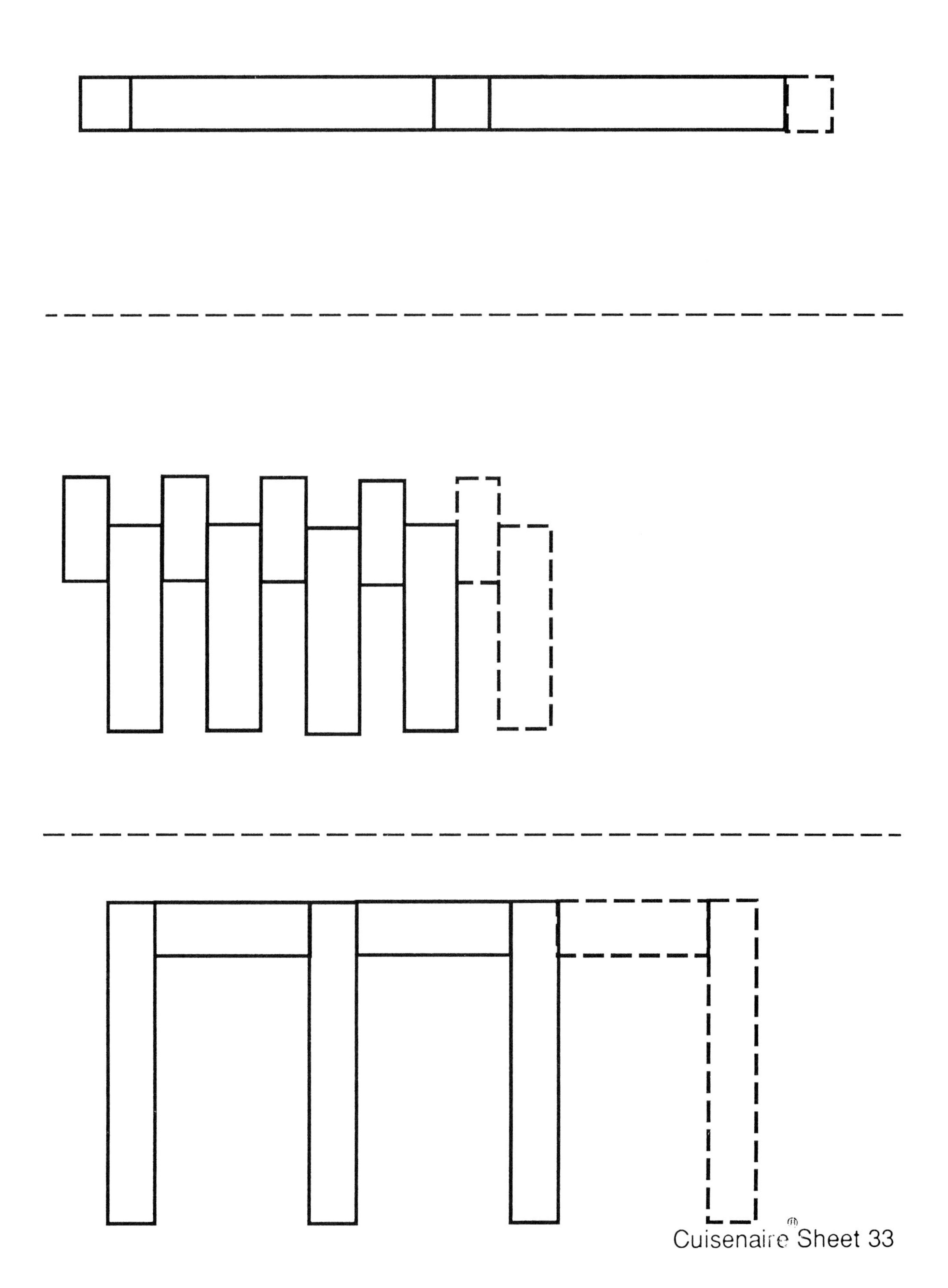
Cuisenaire® Sheet 33

STORY

It is hoped that the story which follows will be enjoyed for its own sake as well as being a source of Mathematical ideas.

THE PROBLEM WITH GRUMBLIES

When Claire woke that morning she knew there was an adventure about. She could feel it in the air. The birds were singing about it, the cat was chasing her tail. Even the dog who was usually too tired to open more than one eye was bouncing round the kitchen. In fact, the feeling was everywhere. No- one else noticed it. Her brother Jonathan kicked her for no reason and when she tried to tell her mum, she didn't really listen. She just told her to finish her toast and began clearing the table. Claire went upstairs to clean her teeth and she felt it again, and knew for certain that an adventure was really going to happen.
The question was 'WHEN?'

The street which Jonathan and Claire lived in looked like any other street in their town, but to them it was very special. So was their school which was just round the corner, and could be seen from Jonathan's bedroom window if you stood on a chair. They loved their teacher, Mrs. Spencer she had a twinkle which made all the things they had to learn special as well. Every day they walked to school.

Today they were hurrying along the road because they were late. School would have started by the time they got there. Walking as fast as they could they rushed round the corner to the school gate.
It wasn't there!
In fact nothing was there, no gate, no school, no playground, no children, NOTHING, just a wide open space with a box in the middle of it.

"What shall we do?" said Claire.
"Let's go home," Jonathan answered.
"What's in the box?" asked Claire.
"How should I know," Jonathan replied.
"Lets go and look," suggested Claire.
"Lets go home, I'm scared." said Jonathan.

Claire certainly didn't feel scared. She knew that this was the adventure she had been expecting and she wanted to look in that box.

"You can go home if you want," she said to Jonathan "I'm not!"
She began walking towards the box. Jonathan followed slowly.

The box was black and made of tin. It had a handle on the top, a lock with a key in it and a label which said, "OPEN ME PLEASE.".

Their mummy and daddy had always told them to leave strange things alone. This, however, was a crisis. They'had lost their school. Something had to be done! Claire bent down, turned the key, lifted the lid and looked inside. What do you think she saw?

Inside there was a strange old key and two pieces of paper, which she picked up and looked at. One was a map and the other a letter. "What does it say?" asked Jonathan who had appeared beside her. "I don't know, silly ," she said "I can't read.". Jonathan, who could, snatched it out of her hand and began reading.

"This letter is from the people who live in Grumblyland. We haven't washed our necks or ears for years. Our school is rotten so we have taken yours. If you want it back you'll have to come and get it. We didn't want to tell you where it had gone but our teacher made us. There's a map in the box to help you. At the places marked with a cross there are problems for you to solve. If you want your school back you'll have to solve them. We bet you can't! The first thing you must do is to find the entrance to Grumblyland. It is beneath the old elm tree in the park. Take the key with you. Climb through and follow the Pattern Block Road."

From Grumblies

What should they do? Go home? Pester the Police? Alert the Army? Ring the Red Cross? Fetch the Fire Brigade? Or perhaps they should go and look for the old tree! What would you do?

They decided to go and look for the tree because they couldn't remember one being there before. Down the road, past the shop and into the park they went and there it was! The tree was by the path. They knew it was the right tree, because in the middle of the trunk was a keyhole. Have you ever seen a tree with a keyhole in the middle of it? Neither had Jonathan or Claire!

Claire slowly pushed the key into the hole and turned it. Nothing happened at first. Then, slowly and silently the tree leaned right over. Underneath, the children could see some stone steps leading downwards. As they stared they thought there was a voice calling them to step through the hole. They couldn't see anything dangerous so looking at each other, taking a deep breath and holding hands they crept down.

In front of them was a road made from coloured stones of different shapes. There were yellow hexagons, red trapezoids, green triangles and blue diamonds stretching into the distance. At the side was a sign like a pointing finger, on the sign was written 'GRUMBLYTOWN' and underneath in large red letters it said "If you leave at playtime you'll be home for dinner."

Note to the teacher:
1. Discussion on time, how long is it to dinner time and what they could do in that time might be interesting.
2. Story sheet 1 is designed to be completed with Pattern Blocks, coloured in, used for creative writing or whatever appeals. The pupils should be encouraged to continue the Pattern Block Road off the page or better still, design one of their own.

The ground was covered with grass. Grass that was long enough to hide in but short enough to run in. It stretched as far as they could see with tall trees dotted here and there. Through the middle of it ran the Pattern Block Road and in the distance they could see some houses. Claire looked at the map and saw a little town marked with a cross. Jonathan slowly read out
"Middleville - half way to Grumblyland. I bet we'll find the first problem there!" he shouted.
"Well let's go and solve it." said Claire.

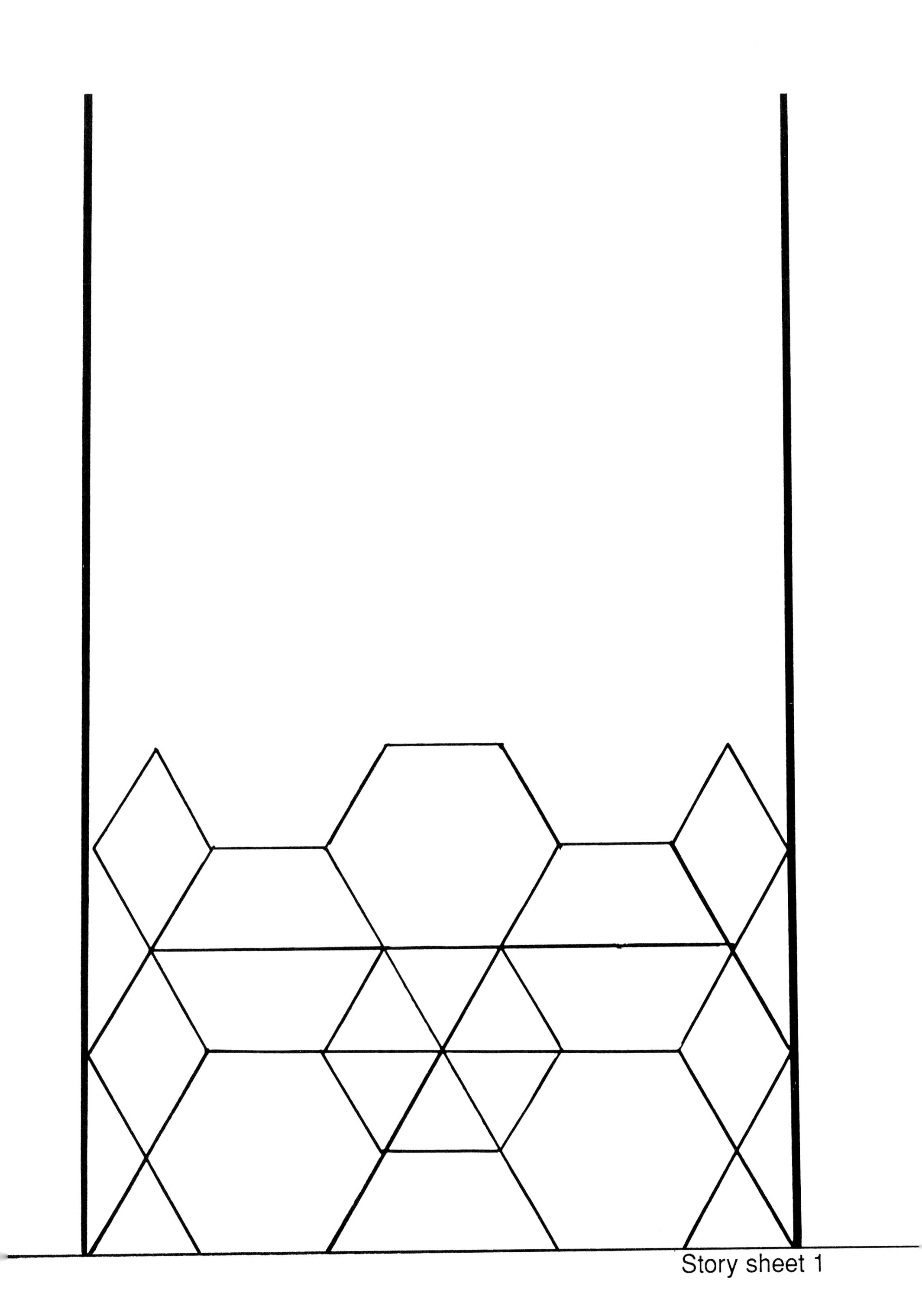
Story sheet 1

They began to walk down the road, (making sure not to step on any lines). As they walked they counted out loud every time they touched one of the hexagons. As Claire could only count to 20, they counted to 20 and then started again. After they had done that 10 times the houses looked a lot nearer.

Note to the teacher:
Ask the children :-
i. to estimate how long it would take to count to twenty.
ii. how long it would take to count to twenty ten times.
iii. if they know the number Claire and Jonathan would have reached if Claire could count beyond twenty (a calculator might help).
iv. how many times they can count to twenty on their way to and from school or any other familiar place.

As they got nearer the houses they could hear shouting. A crowd of little people dressed in funny clothes were standing by a river. Some looked confused, some shouted, others waved their arms about, some pointed their fingers at a man dressed like a ships captain. He was shouting back and pointing at a large wooden box that was marked 'BOAT OR BOATS'.

"I know how to sail boats not make them!" he yelled.
"That's no use to us. We want to get across the river!" the people shouted at him.
"Can we help?" asked Claire.
Everything went quiet and the people turned and stared.
"Funny looking pair," someone said.
"Perhaps they're from the Boat Company." suggested the Captain.
"Do you know how to make boats?"
"Are you from the Boat Company?"
"This letter must be yours ." he said thrusting a large yellow envelope at them. Jonathan opened it up and began to read.

It said,

"TO GET TO GRUMBLYTOWN BUILD A BOAT TO CROSS
THIS CROCODILE- INFESTED RIVER.".

They certainly did have problems:- angry people, a crocodile infested river, a mysterious wooden box and a boat to build.

Claire looked at Jonathan and Jonathan looked at Claire.
"What does it say?" demanded the Captain.
"It says we've got to cross the crocodile infested river," said Jonathan.
"Then you'll have to build the boat," interrupted the Captain.
"Let's look in the box," said Claire.

They all moved over to the box and began opening it.
What do you think was inside?

Let the children guess.

Inside there were giant bricks.

Jonathan and Claire unloaded the bricks and began making a boat. When it was finished they stood back, looked at it proudly and waited for a word of praise from the Captain or the crowd of Grumblylanders.
"That's no good," said the Captain.
"Why not?" snapped Claire.
"Because we can't all get in it." said the Captain.
Murmurs of agreement came from the crowd.
"It's the biggest one we could make," said Jonathan "there are no more bricks."
"You didn't use them properly," said the Captain "I bet I can make a bigger one!"
"Go on then." they said.
Without another word the Captain took their boat apart and began making another. After much huffing and puffing he finished and it did not seem any bigger than theirs had been. They couldn't compare them because theirs had been broken up , so they decided to say nothing.

Note to the teacher:
Any building blocks can be used, for example interlocking cubes can be made leakproof with plastic film. Plasticine or clay are just as good as the children have the initial problem of making the right shape to get them to float. This can be followed by tests of buoyancy.
Having made the boats:-
1. Discuss what Jonathan and Claire might have done to show that the Captain's boat was no bigger than theirs.
2. Have a competition to see who can make the biggest boat with a given amount of material. This will be a chance to discuss comparison between deepest, largest, carries the most Grumblies etc.

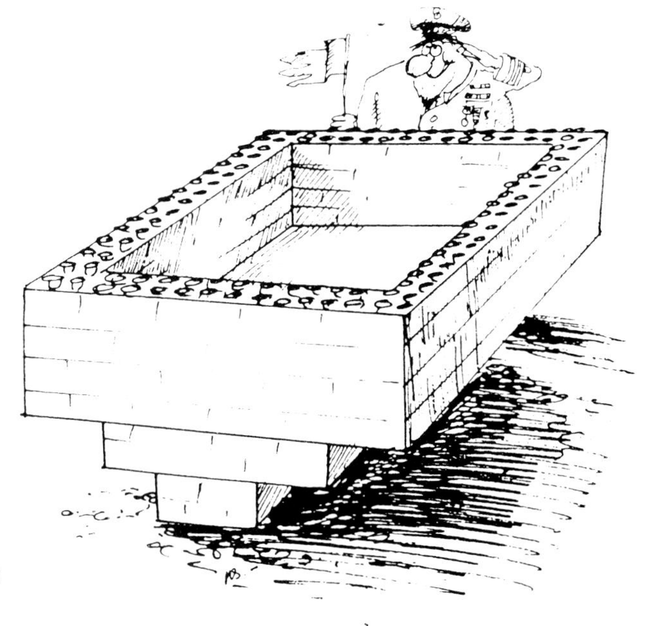

"We can't all get in there." said the large lady with the pointed hat.
"That's easy!" said the Captain, (he was used to this sort of problem) "we'll take a few at a time, and as you two were last here you can go last." he said to Jonathan and Claire.
"How many trips will that be?" Claire wondered.
Can you work it out for her?
(Use story sheet 2)

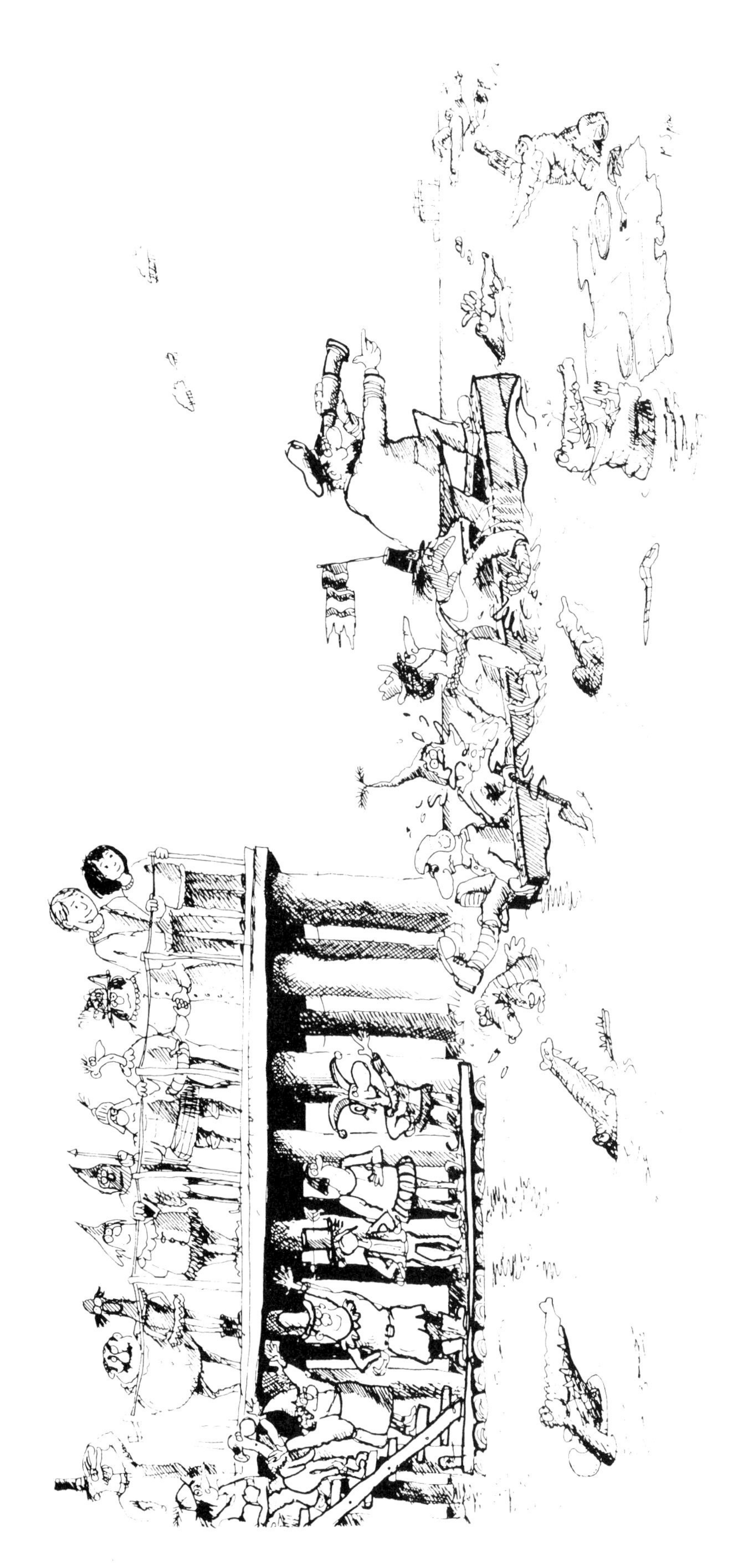

Story sheet 2

ACROSS THE RIVER

Finally everyone was across and Jonathan and Claire began walking along The Pattern Block Road. Some delicious smells made them realise that they were rather hungry and a bit thirsty. As if by magic a cafe appeared at the road side. When they looked at the map, the place was marked with a cross.

"This is where the next problem is!" they both shouted.

They opened the cafe door and walked into a warm cosy room. In front of them was a brightly lit glass counter. Inside it was the most amazing display of food they had ever seen. There were fruit cakes, sponge cakes, cakes with icing, cakes without icing, cakes with cherries and even cakes with things on they had never seen before. Around them were:- apples, pears, grapes, oranges, bananas, in fact every type of fruit they had ever heard or dreamed of. Next to all this was the ice cream. Pots and pots, full of every colour and flavour imaginable. There was red ice-cream, green ice-cream, spotted blue and yellow ice-cream, chocolate ice-cream with nuts and coconut and many more that were completely new to Jonathan and Claire. It was the most magnificent display of food they had ever seen. Everything was named and on the wall was a strange price list.

"Have you got any money?" asked Claire.
"I've got our dinner money" said Jonathan.

"Can I help you?" asked a cheerful-looking Grumblylander from behind the counter who was dressed completely in red, except for a tall white hat on his head.
"Yes please..... but there's so much to choose from, we'll have to think for a minute!"
"That's fine, take as long as you like." he said.
When they had chosen their food and it was served he said,
"Now give me some money please."
"How much?" they asked.
"I don't know, you'll have to work it out for me," he replied.
"It's quite easy, you write down what you've bought, look at the alphabet list and find out how much each letter costs, add it all up and that's what you pay me."
When they added it up they found they had just enough money.
What do you think they bought?

Note to the teacher:-
Use Story sheet 3.
Discuss with the children the price of a meal. Ask them either to spend the same amount of money on items from the shop or choose what they would like and find out how much they cost. We have left the currency off the price list. This can be added if required.

What would you choose?

a- 1	g- 2	m- 3	s- 4
b- 2	h- 3	n- 4	t- 5
c- 3	i- 4	o- 5	u- 1
d- 4	j- 5	p- 1	v- 2
e- 5	k- 1	q- 2	w- 3
f- 1	l- 2	r- 3	y- 4

Story sheet 3

When they were full, they thanked the man in the shop and went on their way to Grumblytown. In the distance they could see some strange-looking posts. As they got closer they realised that they were, trees without branches. The branches were lying in a pile on the ground. A crowd of Grumblylanders were arguing nearby.
"What's the matter?" Jonathan asked.
Claire looked at the map and sure enough there was another cross.
"Its another problem to solve." she whispered. "Can we help?" she asked the Grumblylanders.
"We want to go through the wood but we don't know what to do about the mad magnificent mathematical monster that walks backwards and forwards through the trees." one of the Grumblylanders said.
"What's one of those?" asked Claire.
"It's a wicked-looking creature that has 5 of this, 4 of that, 3 things, 2 widgets and a whatsit !"

Note to the teacher:-
Give the children an opportunity to draw the monster.

"We could climb a tree but all the branches are on the ground. These are special trees, you can stick branches in the holes in the trunks. All the trees must have the same number of branches, otherwise they fall out. Then we can climb the tree, sit on the branch and we are safe from the monster."
"That's a tricky problem," said Claire.
"How many people can go on each branch?" Jonathan asked.
"Only one!" said a man with spotted tights and baggy sweater.

Note to the teacher:
In Story sheet 4 the number of Grumblylanders (15) is exactly divisible by the number of trees (5) - so that each tree should have three branches. Teachers may wish to change the problem and use different numbers .

Claire counted the trees, the branches and the Grumblypeople.
"I know," she said, "we'll have a branch each and when we see the mad magnificent mathematical monster, we'll stick the branch in the nearest tree and climb on it."
"How many people will there be in every tree?" asked Jonathan.

Can you help them?

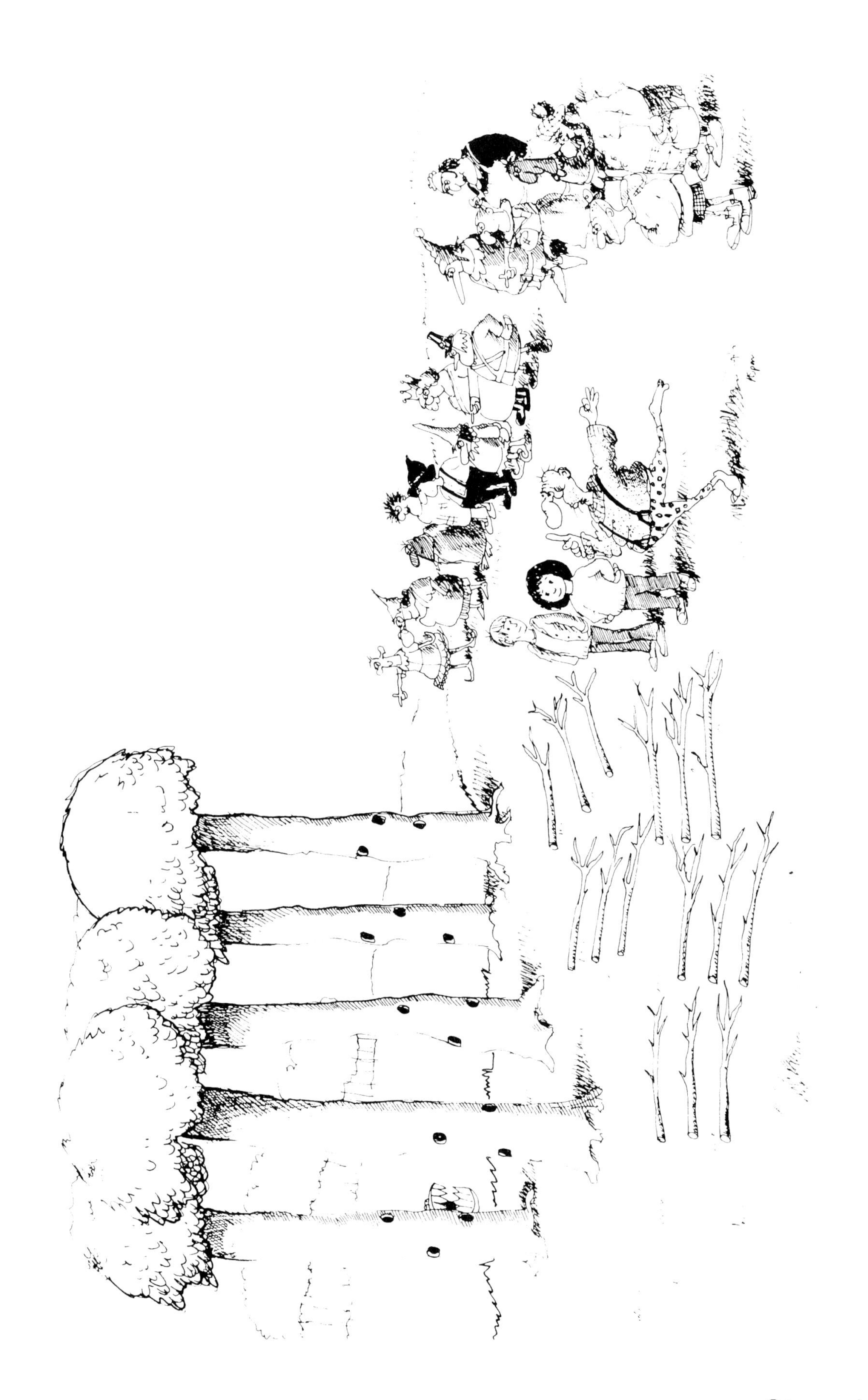

Story sheet 4

When they had all gone through the wood safely, the houses of Grumblytown could be seen in the distance. In the middle of the houses they could see their school. The sight of it made them feel rather excited and they began to run as fast as they could. The school was just as it had been yesterday. The playground was empty and they could hear childrens' voices. They walked through the gates, across the playground and into the building. Their classroom was at the end of the corridor. All the classes they passed were absolutely jam-packed with children. Some they knew, but the rest were Grumbly-children.

They knocked on the door and walked in.
"Thank goodness you've come!" cried their teacher who was sitting at her desk looking very worried.
"I have all these children to teach and they are very naughty. They say that the only way we can leave here is if we say this magic spell properly. I've said it lots of times but it doesn't make any sense to me! When we read it out those Grumbly-children all burst out laughing."
"Let me see" said Jonathan, feeling very grown-up.
Mrs Spencer gave him a piece of paper with writing on.
Jonathan read it out. She was right. It did not make sense and when he was finished all the Grumbly-children started laughing.
"There has to be something we've not spotted," he muttered, and as he said that a smile came over his face.
"I know what it is," he said and began reading it out differently.
What had he seen?

Note to the teacher:-
Two messages for children of different reading abilities are on Story sheet 5. When they decode them it can be fun to get them to rearrange messages of their own as puzzles for others in the class.

After Jonathan read the message properly there was a loud bang and ......
Where do you think he was?

Note to the teacher:-
Instead of our ending, why not let the children write their own endings?

Jonathan was walking down the road with Claire, they were going to school. Down the hill and round the corner and what do you think they saw?

All	to	say	and	will
you	do	this	home	go
have	is	properly	you	quickly

emoh su ekaT

Story sheet 5

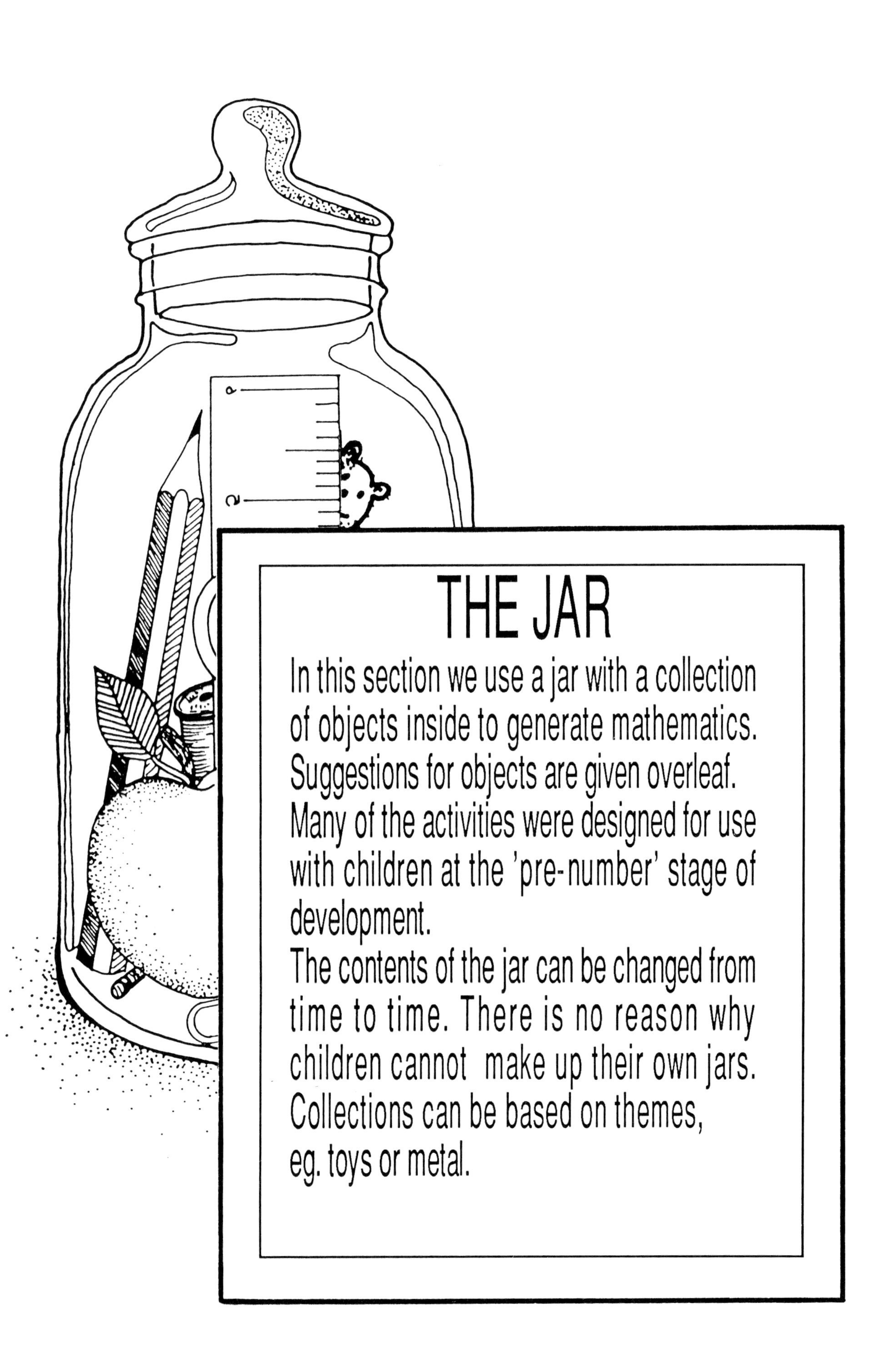

THE JAR

In this section we use a jar with a collection of objects inside to generate mathematics. Suggestions for objects are given overleaf.

Many of the activities were designed for use with children at the 'pre-number' stage of development.

The contents of the jar can be changed from time to time. There is no reason why children cannot make up their own jars. Collections can be based on themes, eg. toys or metal.

For example:-

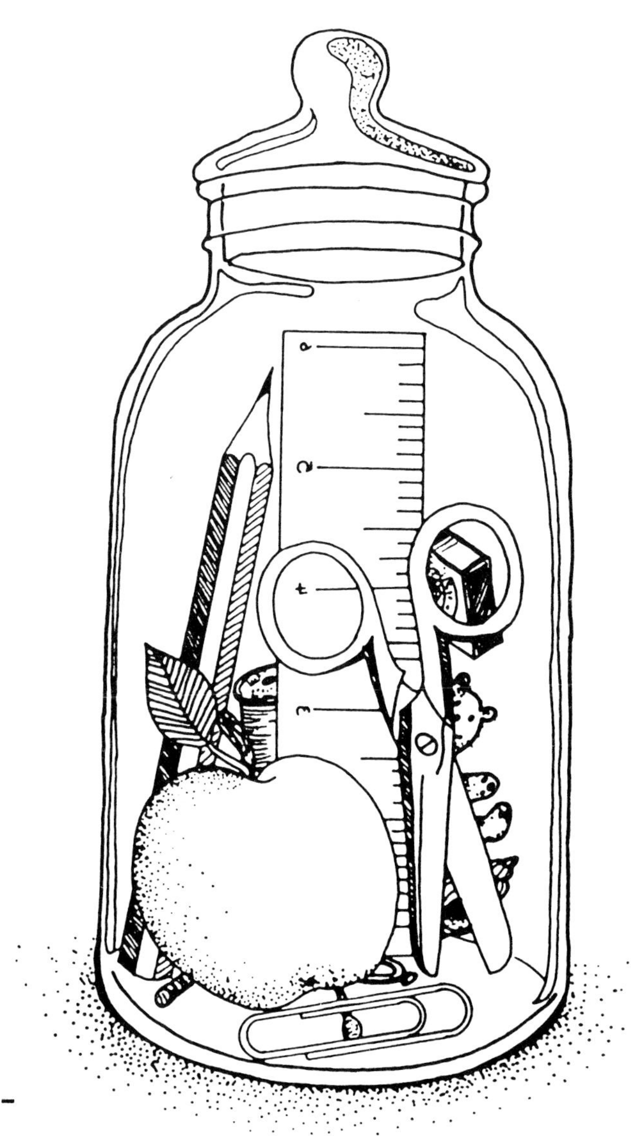

Possible objects:-

pencil
pen
paintbrush
apple
paperclip
matchstick
matchbox
lego or duplo
peg
cotton reel
conker
acorn
leaf
safety pin
key
cube
shape
glove
nut
model cars
nail
feather
lid
notepad
tissue
small teddy
pencil sharpener
spoon
eraser
flower
feather
badge
bracelet
ring
necklace
lump of plasticine
elephant
pipe cleaner
pebble
straw
shell
spark plug
fuse
cotton bud
belt
farmyard animals
shape
ball

Jar Sheet 1

TELL ME ABOUT IT

A game for 2 - 6 players, or teams.

You need:-

7 or 8 interesting objects in an interesting container.

****

These activities develop the skills of sorting, classifying, observing, sharing ideas and learning with others.

****

Introduction

Ask the children to identify the properties of the objects.
For example:-

"What colour is this?"
"What shapes can you see in the scissors?"
"What might this box be used for?"
"What is it made of?"
"What do you think this will feel like?"
"How much do you think it would cost?"

The game can end here or continue with another object. When play has ended children find out who has won, by counting or comparing their counters. For children who cannot count there should be considerable encouragement to help them invent their own methods of comparison. These may include using a balance, stacking, placing them on a number line or matching.

To Play :- for 2 teams or 2 players.

- Take one object out of the jar.

- In turns, each child or team thinks of something different to say about it, if they succeed they collect a cube or a counter.

- Play continues until no-one can think of anything new to say.

Extension

2 objects are taken from the jar.
Children think of common attributes e.g. "They are both blue".

PAIRS

A game to follow 'Tell me about it'.

You need:- 7 or 8 interesting objects in an interesting container
This game can be played either,

i) with 2 teams or players
or
ii) with a "leader" playing with a group of children.

Children take it in turns identifying pairs of objects with a common attribute.
e.g. The scissors and paperclip are both metal.

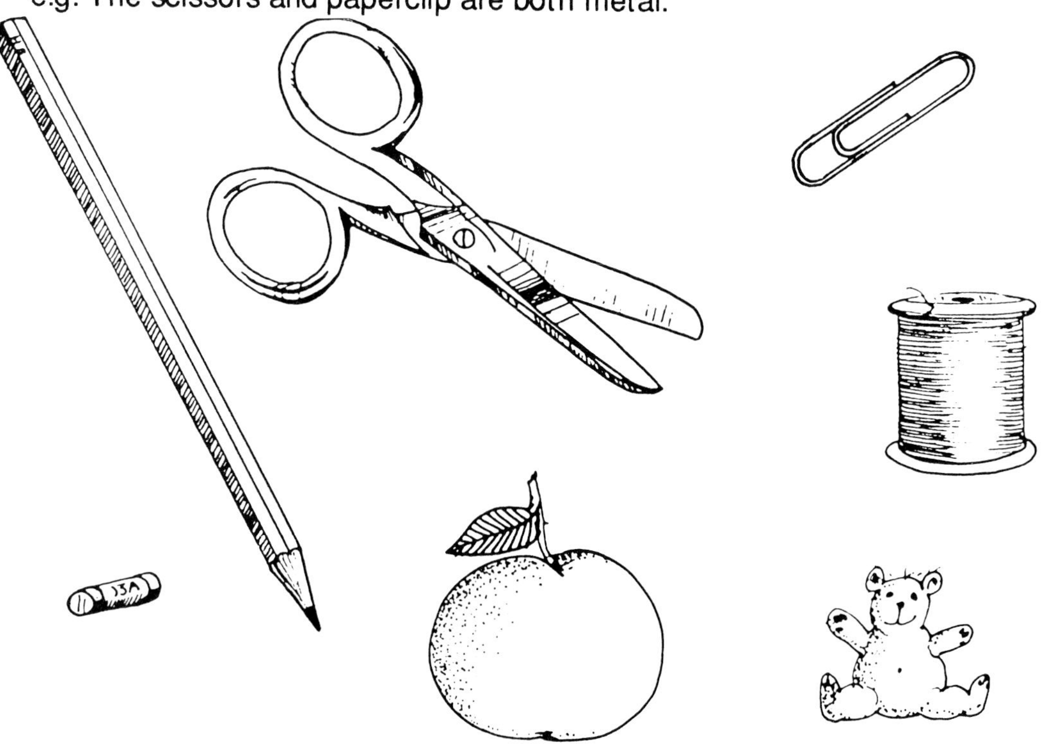

To play:-

- The teacher selects 2 objects with a common attribute.
- Children take it in turns to 'guess' what that common attribute is.
- If the guess is correct the child collects a cube or counter.
- Another 2 objects are selected.
- After 10 rounds, the child with the most cubes or counters wins the game.

Variations

- Children choose the objects.

- Play as a team game.
  The team which thinks of the most common attributes wins the game.

TEN CUBES

A game for 2 - 6 players or teams.

You need:- 7 or 8 interesting objects in an interesting container.

This game is similar to the traditional 20 questions. The simplest way to introduce it is for the teacher to play against the class.

- The teacher draws a picture of an object on a piece of paper as a check.
- The opposing players have ten cubes or counters.
- They take it in turns to ask questions which will help them to identify the chosen object. eg. "Is it plastic?"
  They can only be answered by "Yes" or "No".
- Each time a question is asked a cube or counter is taken away.
- If opponents guess correctly before the cubes run out, the remainder are kept as a running total. The child or team with the highest score after 6 rounds wins the game.

ARE THERE ENOUGH?

****

This is a matching activity to develop conservation of number

****

You need:- A collection of interesting objects in an interesting container.

The Problem

"Are there enough objects in the jar for every member of the group to have one each?"

Discuss with the children how they can find out. Try some of their suggestions. For example, they may decide to count the objects or to give them out. Encourage them to decide on the most efficient (successful) method, and use it with other collections.

Extensions

- Can each member of the group have 2 each?
- Use a jar of coloured cubes and ask the same questions.
- Take a jug of water, can each member of the group have a glass each?
- How many children can have a drink from the jug?
- If everyone has a glass of water, and it is poured into the jug, how far up the jar will the water go?

COMPARISON

Jar Sheets 2-4

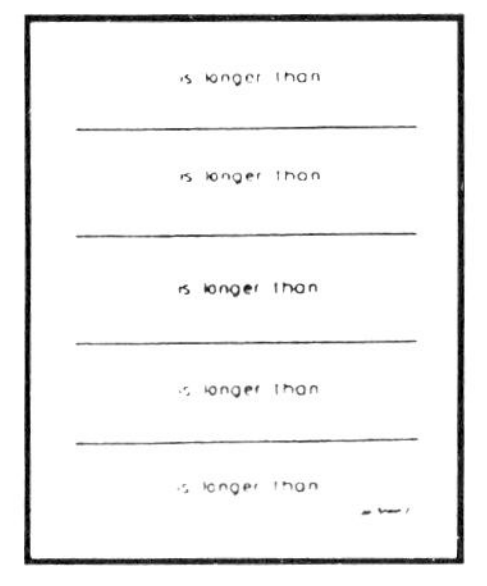

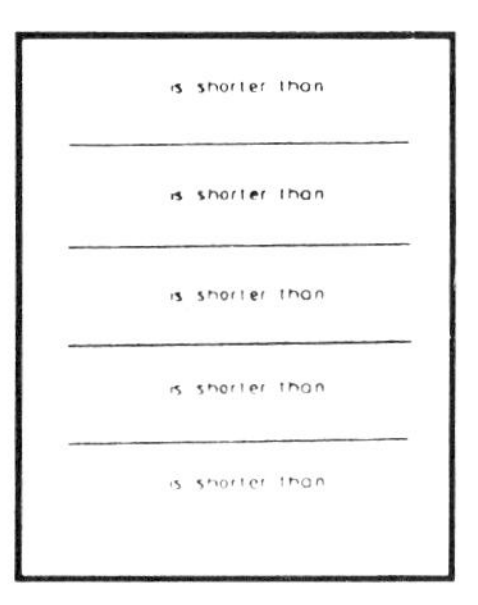

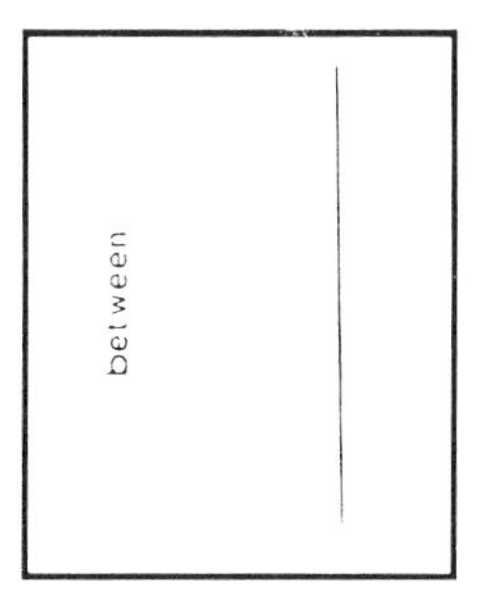

You need:- a jar with 10 objects which are dissimilar in length.

Children choose pairs of objects from the jar and compare their lengths.
They record the activity by either using jar sheets 2 or 3 as mats and placing the objects on, or by drawing pictures of their results.

Extension
Make sure that 2 or more objects in the jar are the same length. Use either sheet 2 or 3 and ask them to find a place on the sheet for each object.

The Between Game

This game or activity can be directed by the teacher or a child.

To Play:-

- Take two objects from the jar and place them at either end of the game board (jar sheet 4).
- The first player/team chooses an object from the jar which they believe to be between these in length. They place it between the two objects which are already on the board.
- If they are correct, they collect a counter.
- The objects are placed back in the jar, and a similar challenge is set for the next player/team.
- The winning player/team is the one who has collected the most counters.

Extension

Ask the children to order more than three objects according to their length.

is longer than

is longer than

is longer than

is longer than

is longer than

Jar Sheet 2

is shorter than

---

is shorter than

---

is shorter than

---

is shorter than

---

is shorter than

Jar Sheet 3

between

Jar Sheet 4

WILL IT FIT?

****

This activity gives children experience in estimation of length, and provides a wonderful opportunity for discussion.

****

You need:- 7 or 8 interesting objects in an interesting container.

Introduction

The children imagine that all the objects in the jar are placed end to end.

How far do they think the line will reach?
Will it reach -

a) across the table?
b) from the beginning to the end of the shelf?
c) the length of somebody's arm?

Give the children a chance to refine their estimate when there are only two more to be placed.

Extensions

1. Find one object that is about the same length as a pencil.

2. Find two objects which together are about the same length as a pencil.

3. Find ten objects which together are about the same length as a pencil.

4. Find four objects which are about the same length as 3 yellow Cuisenaire rods in a train.

5. Children invent four challenges of their own.

6. How can the door be measured?

7. Use rods to find the approximate length of each object.

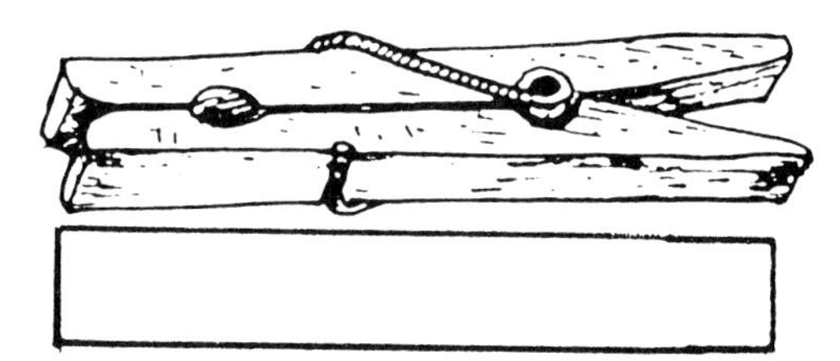

8. Repeat this activity using interlocking cubes, buttons etc.

9. Find the total length of the line of objects in cubes or rods.

MATCHING

Make a number line using Jar sheet 5 or 6.

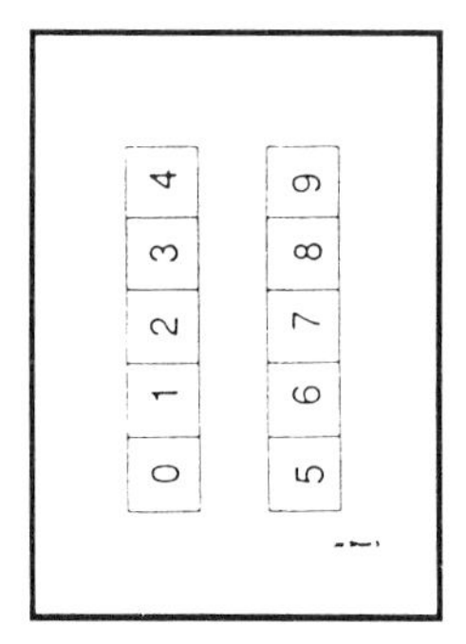

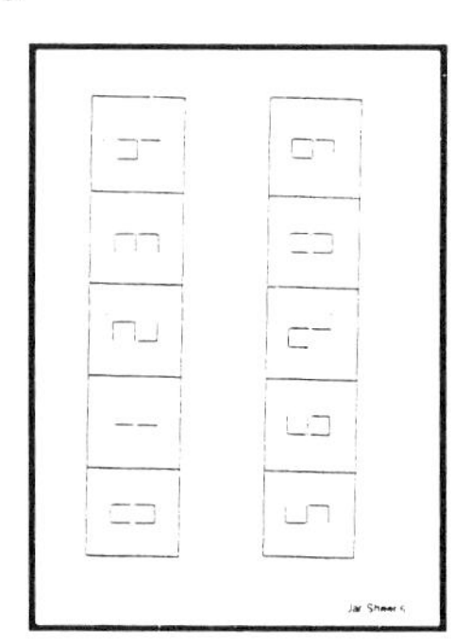

****

This activity gives children experience in estimation and counting.

****

You need:- 7or 8 interesting objects in an interesting container.

- Estimate how many objects there are in the jar.
- The children record their estimates.
- Count the objects by matching them with one of the number lines. (Jar sheets 5 and 6)
- Discuss how accurate their estimates were.

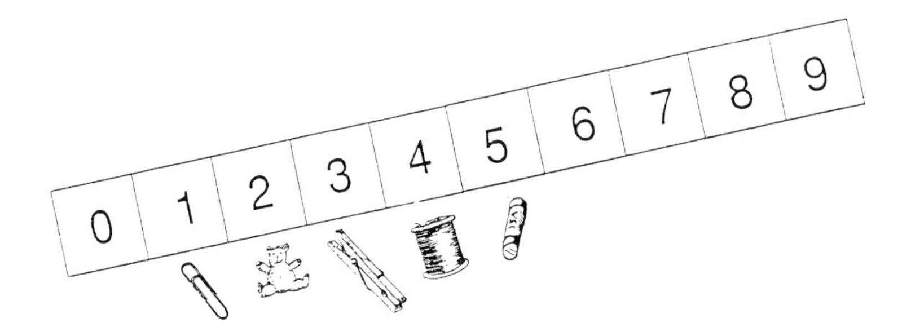

Extensions

1. Find out more about the number of objects in the jar. If, for example, the number was 8,

   - Draw a line for 8 children to stand on.
   - Find a box for 8 books.
   - Make a tower with 8 coloured cubes.
   - Ask the children to work out how many objects there would be if there were one more or one less, two more or two less, etc.

2. Compare the contents of two jars of objects.

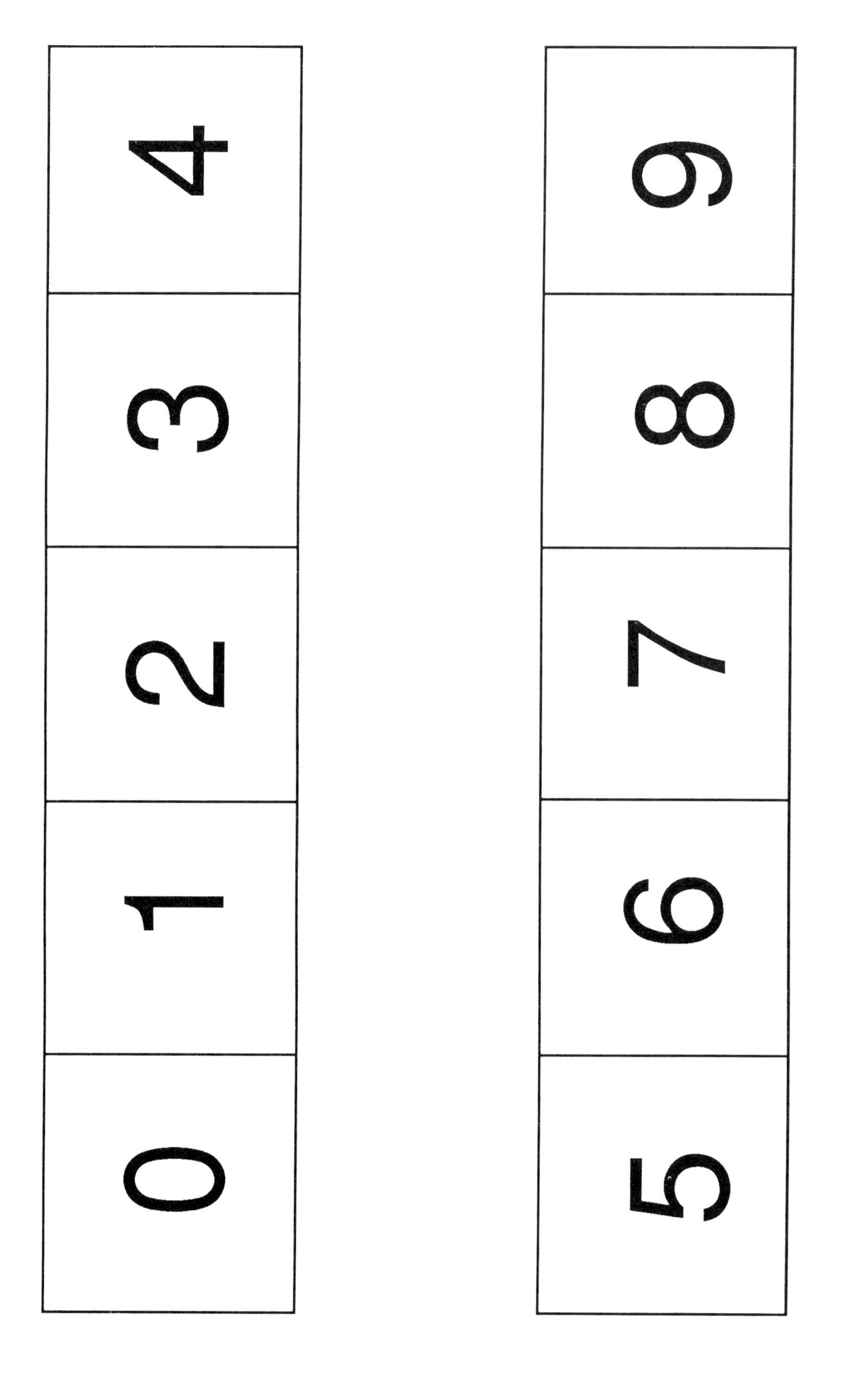

Jar Sheet 5

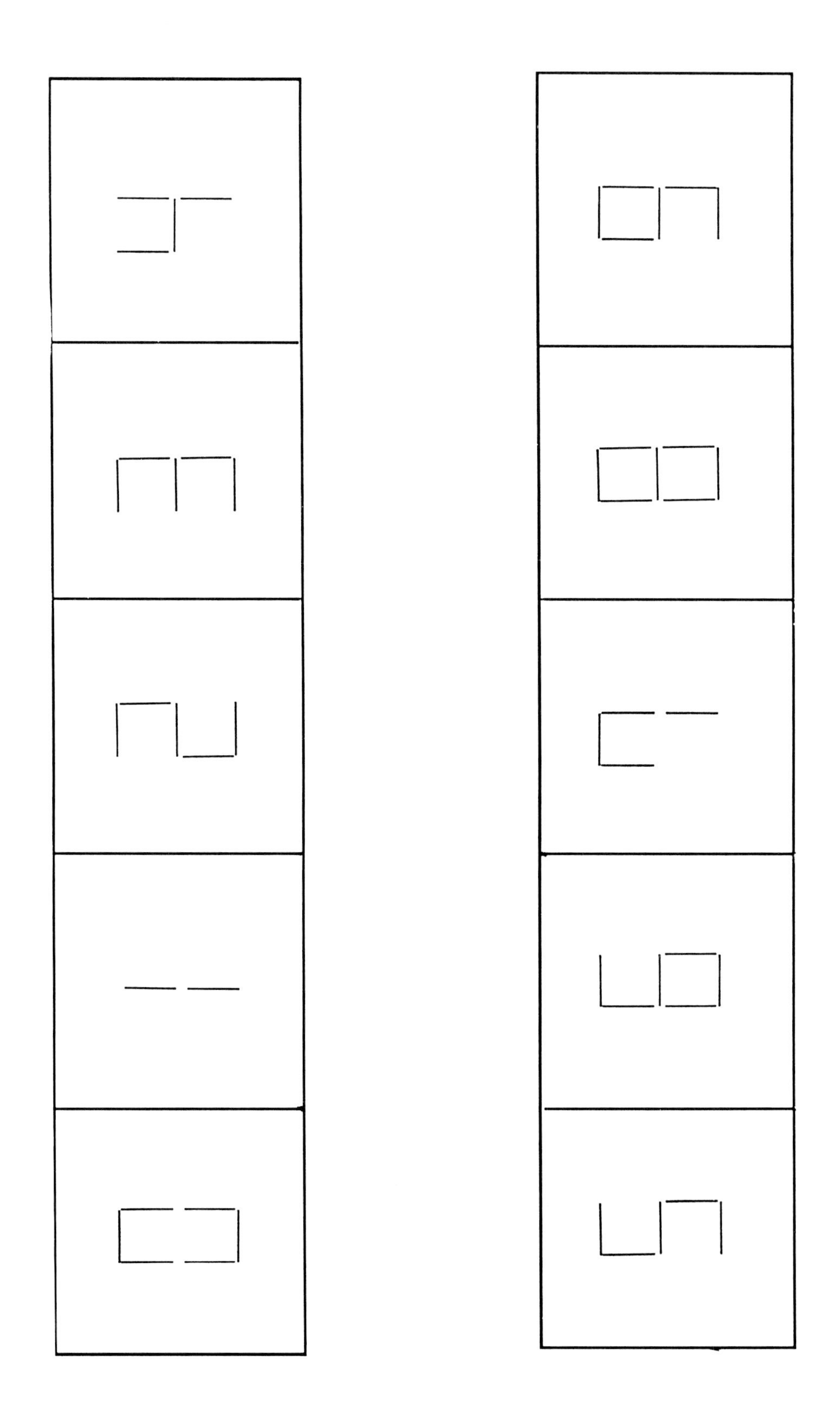

Jar Sheet 6

THE BALANCE GAME

You need:- The jar and a sensitive balance.

To Play

The game is for two players or teams.

- Take it in turns to:-
  - Choose two objects from the jar.
  - Guess what will happen when they are placed on the balance.
  - Check by placing the objects on the balance.
- If the guess was correct, players can collect a cube or counter.
- Choose two more objects and repeat the activity.
- The winners have the most counters at the end of the game.

This game can be played in two ways:-

Either objects are replaced between each turn or not.

BALANCE BOOK

Jar sheets 7 - 9

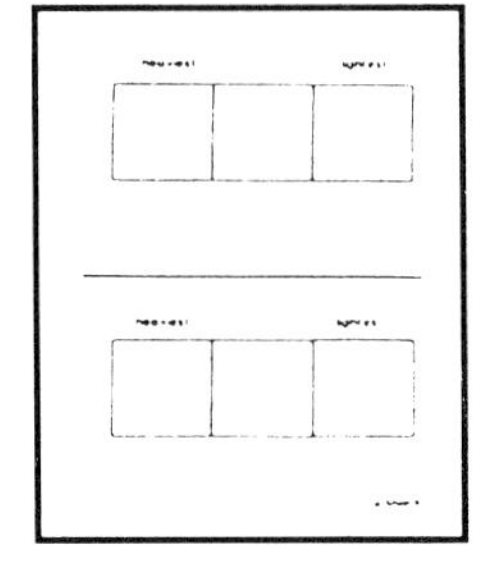

The jar sheets are designed to be made into a Balance Book, which the children can use in a variety of ways, such as.....

1. - Choose two objects from the jar and guess which is the heaviest.
   - Check on the balance and record in the Balance Book

2. - Look at the balance picture.
   - Find, estimate and check two objects that make the balance look the same as the picture.

Extensions

- Take 3 objects and put them in order from lightest to heaviest. Record on jar sheet 9.
- Take another 3 objects and put them in order. Take a fourth and find out where it should go in the order. Continue until all the objects are in order.

This is a very difficult activity, discuss the strategies used by the children.

Balance Book

Name: ______________________

Jar Sheet 7

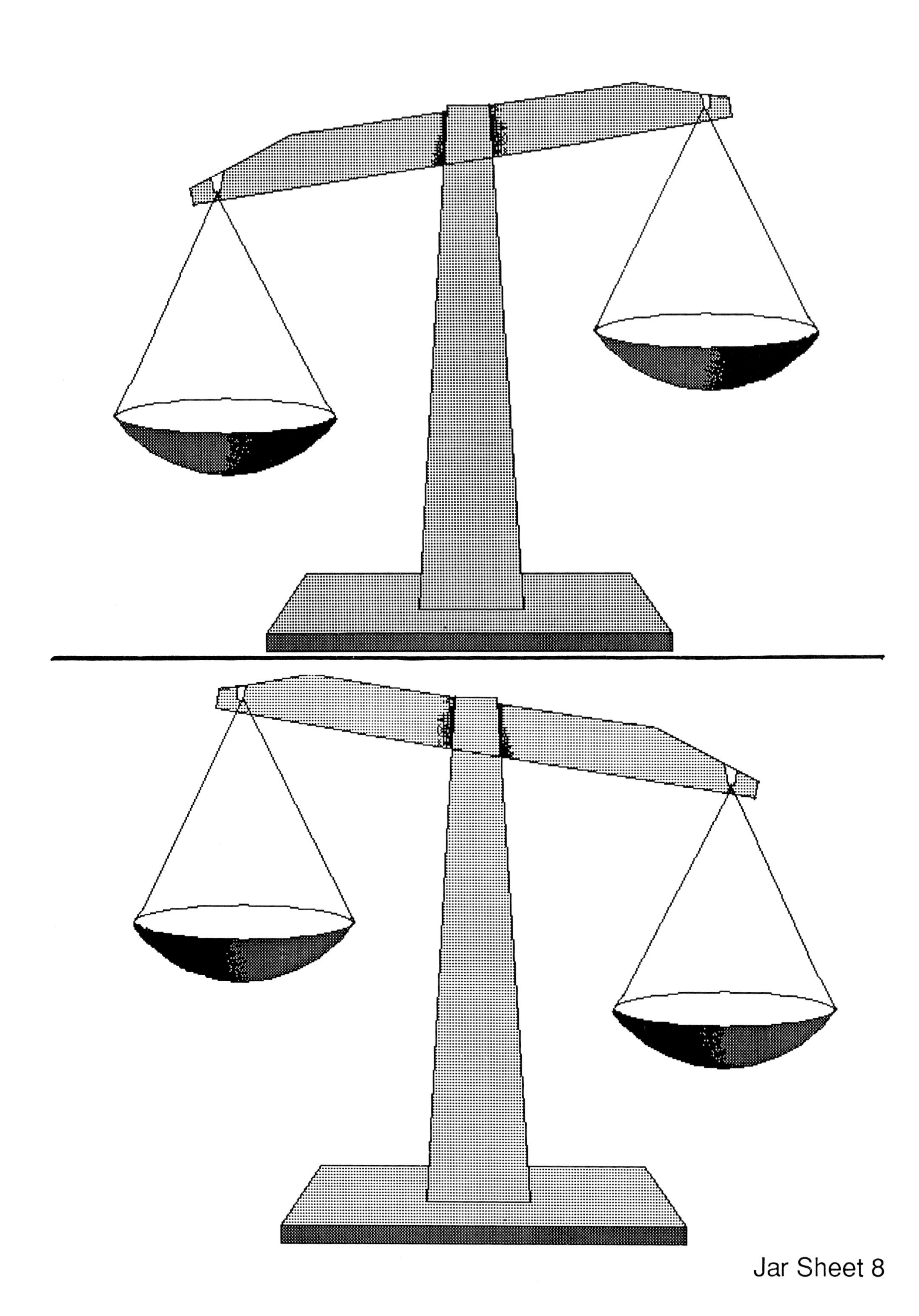

Jar Sheet 8

heaviest		lightest

heaviest		lightest

Jar Sheet 9

BALANCING

1. Ask the children to choose an object, and to make a plasticine model of it.
   Place them both on a balance.
   Discuss what happens.

2. An object is chosen from the jar.
   The children take as many cubes as they think will balance the object.
   The cubes and object are placed on the balance.
   The number of cubes are adjusted until they balance with the object.

   Repeat the activity and encourage the children to predict the outcomes by asking questions like:-

   "Will we need more/less than last time?"
   "Will we need more/less than ten cubes?"

   The results can be recorded in a number of ways. For example:-

......using jar sheet 10

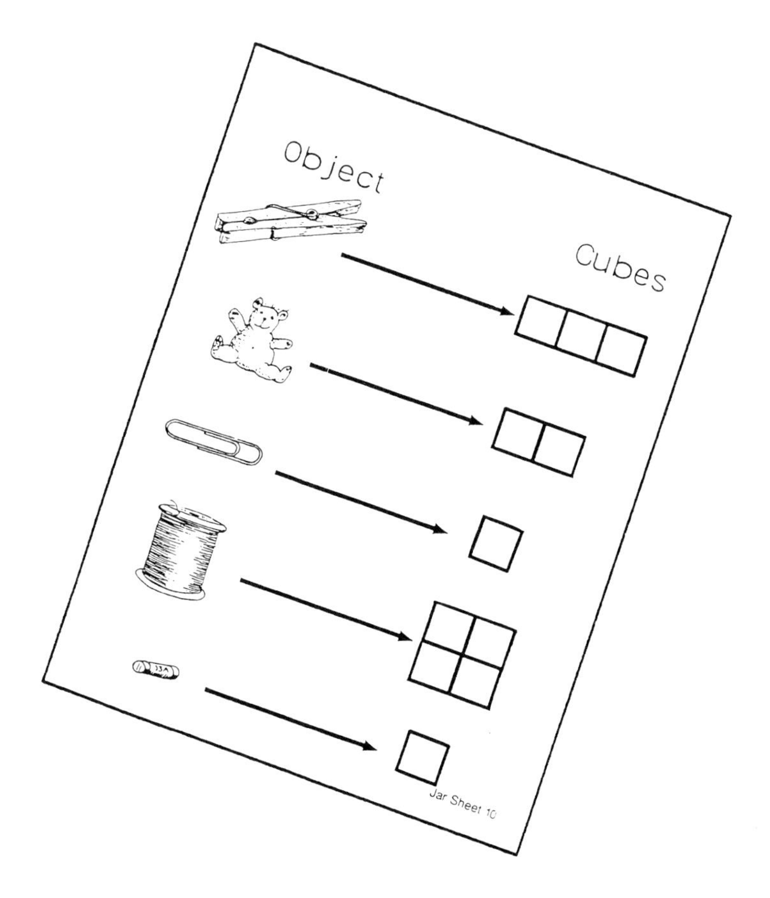

.....or on squared paper.

Object Cubes

Jar Sheet 10

FLOATING AND BOATING

Jar sheet 11

You will need:- The jar and a tank of water.

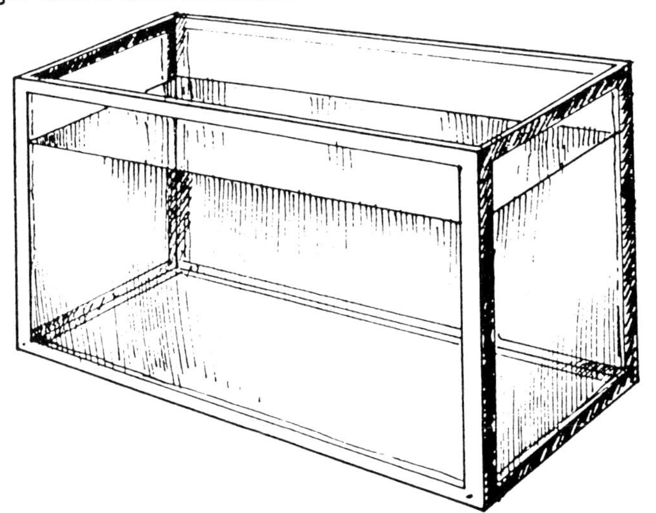

1. Ask the children to find out which objects float and which sink.

   Jar sheet 11 opposite may be used for recording.

2. Give children equal amounts of plasticine to make a boat.
   The following challenges can be set:-

   Who can get the most objects in their boat before it sinks?
   Can you make a boat out of something else? (eg. cubes covered with plastic film)

Object	Float? (Yes or No)

Jar Sheet 11

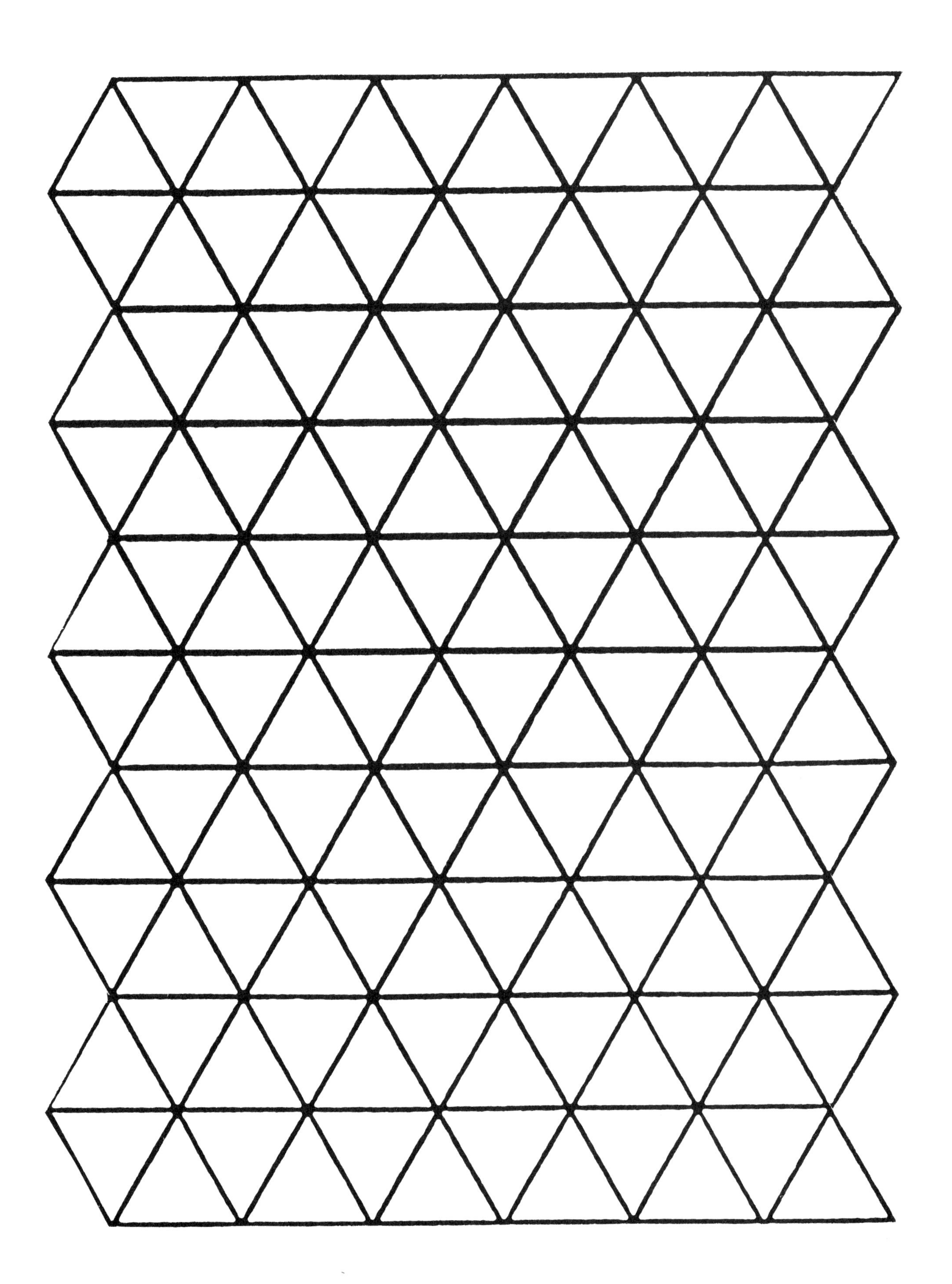

0	1	2	3	4	5	6	7	8	9
10	11	12	13	14	15	16	17	18	19
20	21	22	23	24	25	26	27	28	29
30	31	32	33	34	35	36	37	38	39
40	41	42	43	44	45	46	47	48	49
50	51	52	53	54	55	56	57	58	59
60	61	62	63	64	65	66	67	68	69
70	71	72	73	74	75	76	77	78	79
80	81	82	83	84	85	86	87	88	89
90	91	92	93	94	95	96	97	98	99